Hydrogen

Energy of the Future

Shakti Kumar Pandey

notionpress.com

INDIA • SINGAPORE • MALAYSIA

ISBN
Paperback 979-8-90392-079-2
Hardcase 979-8-90392-080-8

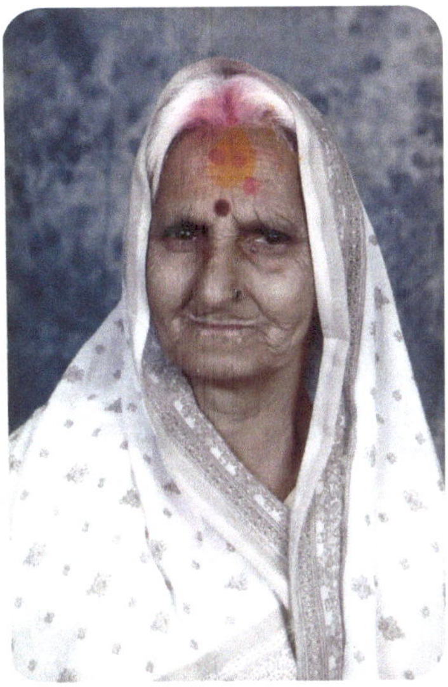

Dedicated to my Mother Parvati Devi.

Contents

Preface

Energy has always shaped the progress of human civilisation. From the mastery of fire to the age of coal, oil, and natural gas, each energy transition has unlocked unprecedented economic growth, while also leaving deep environmental and social footprints. Today, humanity stands at another inflection point. Climate change, air pollution, energy security, and resource constraints are no longer abstract concerns; they are daily realities influencing policy, markets, and lives across the globe. In this context, hydrogen has emerged not as a silver bullet, but as a compelling and versatile piece of the sustainable energy puzzle.

Hydrogen is the most abundant element in the universe, yet paradoxically scarce in its free form on Earth. It must be produced, stored, transported, and converted back into useful energy. These challenges have historically limited its widespread adoption. However, what once seemed impractical is now becoming feasible due to rapid advances in renewable electricity, electrolysis technologies, fuel cells, materials science, and systems integration. As costs fall and performance improves, hydrogen is transitioning from scientific curiosity and niche industrial feedstock to a serious candidate for decarbonising sectors where direct electrification is difficult or impossible.

This book, *Hydrogen – Energy of the Future*, by Mr. Shakti Kumar Pandey, who has more than 3 decades of experience in the oil and gas industry, discusses these challenges in a very cogent manner. This book is written for a diverse audience: policymakers seeking evidence-based guidance, engineers and scientists developing next-generation technologies, industry leaders evaluating strategic investments, students entering the energy field, and informed citizens interested in the future of energy. By examining technical fundamentals, economic realities, environmental impacts, and policy frameworks, the chapters that follow aim to equip readers with the tools to critically assess hydrogen's potential. I strongly recommend this book to anyone who would like to learn about the challenges that must be overcome to increase the production and use of renewable hydrogen.

This book is written at a moment when hydrogen is being rediscovered with new urgency. The current momentum for hydrogen is grounded in necessity. Deep decarbonisation of the global energy system requires solutions for heavy industry, long-distance transport, seasonal energy storage, and high-temperature heat. Batteries and direct electrification will dominate many applications, but they are not sufficient on their own. Hydrogen, along with its derivatives such as ammonia, methanol, and synthetic fuels, offers pathways to address these hard-to-abate sectors.

Hydrogen is not inherently "green" or sustainable. Its climate impact depends entirely on how it is produced. Today, most hydrogen is made from fossil fuels, particularly natural gas, resulting in significant carbon dioxide emissions. This reality underscores a central theme of this book: hydrogen is an energy carrier, not an energy source. Its sustainability hinges on coupling its production to low-carbon or zero-carbon energy systems—most notably renewable electricity. Understanding this distinction is critical to making informed decisions about hydrogen's role in future energy systems.

The purpose of this book is to provide clarity in a field often marked by hype, oversimplification, and polarised debate. A sustainable hydrogen economy is not merely a technological challenge—it is a systems challenge. Producing green hydrogen at scale requires massive deployment of renewable energy, careful water management, and resilient supply chains for critical materials. Transporting and storing hydrogen demands new infrastructure, the repurposing of existing assets, and rigorous safety standards. End-use applications must compete on cost, efficiency, and reliability with incumbent fossil-based solutions. None of these elements can be considered in isolation. Success depends on coordinated action across engineering, economics, regulation, and social acceptance.

While the subject matter is technically complex, the intent is accessibility without oversimplification. Wherever possible, concepts are explained from first principles, and assumptions are made explicit. Readers are encouraged not just to learn what is happening in the hydrogen space, but why.

History offers a useful reminder: energy transitions are slow, uneven, and shaped as much by institutions and politics as by technology. Coal did not disappear with the arrival of oil, nor did oil vanish with the rise of natural gas and renewables. Similarly, hydrogen will not replace all existing fuels. Its adoption will be incremental, sector-specific, and at times contested. Recognising this helps temper unrealistic expectations while still appreciating the transformative potential of well-designed hydrogen systems.

Finally, this book is motivated by a sense of pragmatic optimism. The climate challenge is immense, but so is human ingenuity. Hydrogen will not solve the energy transition alone, but it can play a vital supporting role if deployed thoughtfully and responsibly. The choices made in the coming decade—about technology investment, infrastructure development, and policy design—will determine whether hydrogen becomes a cornerstone of a sustainable energy future or remains a missed opportunity.

The chapters ahead do not promise easy answers, but they aim to illuminate trade-offs, clarify uncertainties, and highlight realistic pathways forward. In doing so, this book hopes to contribute, in its own modest way, to the broader effort of building an energy system that is cleaner, more resilient, and more equitable for generations to come.

Professor **Deepak Kunzru**, FNAE, FNASc
Former Head, Department of Chemical Engineering and Dean, R&D, Indian Institute of Technology Kanpur
Former Dean, Graduate School and Research, Ahmedabad University

Abbreviations

AEM	Anion exchange membrane
ATR	Autothermal reforming
AWE	Alkaline water electrolysis
BOP	Balance-of-plant
CCS	Carbon capture & storage
CCU	Carbon capture and utilisation
CCUS	Carbon capture, utilisation, and storage
CEEW	Council of Energy, Environment, and Water
CSA	Canadian Standard Regulation
DRI	Direct reduction of iron
EDP	Energias de Portugal
EOR	Enhanced oil recovery
EPC	Engineering, procurement, and construction
ETC	Economic Transitions Commission
FEED	Front-end engineering design
FID	Final investment decision
FPG	Fracture pressure gradient
FRP	Fibre-reinforced pipeline
GHG	Green House Gas
GHCI	Green Hydrogen Certification Scheme of India
GoI	Government of India
GTI	Global Technical Institute
GWP	Global Warming Potential
HEFA	Hydro processing of Ester and Acetic Acids
HER	Hydrogen evolution reaction
HESC	Hydrogen Energy Supply Chain
HFC	Hydrogen Fuel Cell
HHV	Higher heating value
HTE	High-temperature electrolysis
HTS	High-temperature shift
ICOS	Integrated Carbon Observation System
IEA	International Energy Agency
IECC	India Energy & Climate Centre
IGCC	Integrated gasification combined cycle

IPCC	Intergovernmental Panel on Climate Change
IREL	Indian Rare Earths Limited
IRENA	International Renewable Energy Association
LCOE	Levelised cost of electricity
LCOH	Levelised cost of hydrogen
LOHC	Liquid organic hydrogen carrier
LTS	Low-temperature shift
ML	Mauna Loa Observatory
MMT	Million metric tonnes
MMTPA	Million tonnes per annum
MNRE	Ministry of New and Renewable Energy
MoPNG	Ministry of Petroleum and Natural Gas
MR	Mixed refrigerants
MSE	Medium and small enterprise
MW	Megawatt
NASA	National Aeronautics and Space Administration
NGHM	National Green Hydrogen Mission
NSG	Nuclear Suppliers Group
OER	Oxygen evolution reaction
OPEC	Organisation of Petroleum Exporting Countries
PEC	Photoelectrochemical
PEM	Proton exchange membrane
PGM	Platinum group metals
PPWS	Put Panels Where It's Sunny
PSA	Pressure-swing adsorption
PV	Photovoltaic
RTC	Round-the-clock
RTI	Research Triangle Institute
RWGS	Reverse Water Gas Shift
SAREP	South Asea Regional Energy Partnership.
SIGHT	Strategic Interventions for Green Hydrogen Transition
SAF	Sustainable Aviation Fuel
SDS	Sustainable development scenario
SMR	Steam Methane Reformation
SNR	Steam naphtha reformation
SOE	Solid oxide electrolysers
SOEC	Solid oxide electrolysis cell
UC	Used cooking oil
UNFCCC	United Nations Framework Convention on Climate Change
WGS	Water gas shift

Introduction

Global industries emit 40 billion tonnes of carbon dioxide annually, mostly from fossil fuel combustion, which is beyond the capacity of the natural carbon cycle to absorb. This results in the accumulation of around half of total industrial carbon dioxide emissions, which contributes to global warming. Global warming is a threat to human survival because it offsets the ecological balance on Earth, a balance between the atmosphere, the hydrosphere, and the biosphere. This, in turn, causes climate change. Climate change brings extreme weather conditions, drought, uneven and excessive rainfall, soil erosion, melting glaciers, sea-level rise, and the migration of humans and animals, threatening their survival.

Reducing and reversing carbon dioxide emission have become a global issue. It has been found that this can only be done by reducing fossil fuel combustion by industries, which emit 80% of global industrial carbon dioxide emissions. Fossil fuels can be replaced to a large extent with renewable energy sources such as solar, wind, hydropower, and biomass. And for that, global efforts are being made to tap the optimal potential of these renewable resources to decarbonise the atmosphere. But these resources alone cannot reduce and reverse the global carbon emissions because of their intermittency, limited availability, and many geographical reasons.

Hydrogen has emerged in a big way to decarbonise the atmosphere, if it is produced from a renewable power source, and no net carbon dioxide is emitted during production, called clean or green hydrogen. Hydrogen emits no carbon dioxide on combustion and is available in abundance on Earth in water form. Two-thirds of the globe is covered with water. Hydrogen, being a strong reducing agent, can break down carbon dioxide in the atmosphere more effectively than any other pathway. It is said that hydrogen has the potential to replace fossil fuels to a very large extent and is termed the Fuel of the Future.

The book discusses where and how clean hydrogen can be sourced, how it can be produced and handled to replace fossil fuels safely and economically, and how it can be made accessible and affordable to the public. R&D, technological advancements to produce clean hydrogen to compete with natural gas, ways to mitigate the risks in handling, and the future prospects of hydrogen as an alternative to fossil fuels have been discussed.

Emphasis has been given to electrolytic technology, which is a compact process and yields green hydrogen with the highest degree of purity and energy efficiency. The electrolytic process is the second most common process after fossil fuels for global hydrogen production, though its share is much lower, 1% as of now. Economics and constraints in bringing electrolytic hydrogen to the energy market have been taken into account.

Different sources of hydrogen, fossil and non-fossil, with various production methodologies, such as electrolytic, thermogenic, biogenic, dark fermentation, storage and transportation, and utilisation in various industries for decarbonisation purposes, have also been discussed. A global review of supply and demand has also been referred to at appropriate places.

In India, hydrogen is classified into grey, blue, and green hydrogen based on associated carbon emissions during production. Grey hydrogen is produced from fossils (high CO_2 emissions), blue from fossil with carbon capture & storage (CCS) facility, and green from renewable energy through water electrolysis (zero/low emissions), as defined by India's Ministry of New and Renewable Energy (MNRE) under its Green Hydrogen Standard and the National Green Hydrogen Mission (NGHM) to promote clean energy. Blue and green hydrogen are said to be clean hydrogen.

Carbon capture, utilisation, and storage technologies to convert grey hydrogen to blue hydrogen, and how best it can compete with natural gas in terms of economics and future energy prospects, have been explained. The future market of green and blue hydrogen, called clean hydrogen, with a timeline, as published by global agencies, and how it is going to shape the hydrogen landscape by the middle of the century, in the global energy market, is also discussed.

Hydrogen fuel cells, the challenges of bringing them to the mobility sector, their advantages and disadvantages, critical metals and their availability that shape the future of renewable energy, and the electrolysis and fuel cell markets have been discussed.

Today, the cost of green hydrogen through the electrolytic process is about 5 times higher than natural gas. There are no or very few infrastructures; only the National Aeronautics and Space Administration (NASA) has a liquid hydrogen transport pipeline available all over the world that can exclusively handle liquid hydrogen as of now. The European Union has started developing an infrastructure for hydrogen handling on a large scale due to the energy crisis that arose because of the Russia-Ukraine war recently, interrupting natural gas supply to them.

Hydrogen cost is high due to the high cost of green power, which is a feedstock for the electrolyser, and the high electrolyser manufacturing cost. These 2 components together share more than 90% of the green hydrogen production cost. The true round-the-clock (RTC) renewable energy cost, which includes factors like storage and grid integration, is higher than the headline prices seen in some solar auctions. Lowering electrolyser cost through indigenous manufacturing and material advancement is crucial for achieving cost parity with hydrogen from fossils. All these aspects of bringing hydrogen into the energy market have been discussed in the book.

Initiatives have been taken by the Indian government to reduce the cost of green hydrogen through policy measures like waivers on inter-state transmission charges and open access charges for renewable energy, as well as financial incentives under the Error! Hyperlink reference not valid.. Other measures include promoting domestic manufacturing of electrolysers, simplifying business operations through single-window clearances, and encouraging research and development. Indian and global efforts being made in these directions have been discussed in the book.

Handling hydrogen is difficult mainly because of its high diffusivity (about 4 times higher than natural gas), which can cause leakages and hydrogen embrittlement in metallic pipelines, and a high flammability range (7 times wider than natural gas), which can cause fire hazard. These 2 characteristics of hydrogen make the handling risky. Global Technical Institute has recommended blending hydrogen into natural gas up to 20%, at a pressure below 86 bar, without major changes to existing natural gas pipelines, and this is being practised in all European nations and in India as well. All these handling issues with hydrogen have been taken into account.

R&D works are going on to improve the energy efficiency of the electrolyser and find suitable ways and means to transport hydrogen safely. Solid oxide electrolysers have been found to yield high efficiency, as high as 90%, but they need high temperatures to operate; therefore, sourcing high-temperature steam from nuclear plants shall be the most suitable proposition. Fibre-reinforced pipeline (FRP) has been found to sustain hydrogen embrittlement up to a limited pressure, but it cannot sustain high temperatures. All these bottlenecks in bringing out clean hydrogen to the energy market are being addressed.

The concept of a hydrogen economy has been discussed, which can become a reality only when green hydrogen will come in energy market to compete with natural gas. For this, green hydrogen must penetrate the mobility sector, which consumes 65-70% of petroleum products globally.

The role and availability of critical metals, including rare earth elements, are taken into account while discussing the production of green hydrogen. The future of green hydrogen largely depends upon the critical metals needed for renewable energy and electrolysers. India is dependent on countries like China, Africa, and Latin America for these critical metals. Rare earth elements (REEs) extraction and processing technology is available to a limited number of nations, with China holding a near-monopoly on advanced refining capabilities and technical expertise. Critical metals for green hydrogen production include platinum group metals (PGMs) like platinum and iridium, used as catalysts in proton exchange membrane (PEM) electrolysers and fuel cells, and nickel, which is a primary material used in alkaline electrolysers. Other critical metals include REEs for magnets such as zirconium, cobalt, and chromium, which are used for strength, corrosion resistance, and other specific applications in various electrolyser technologies. These magnet metals are also required for solar and wind power, which are the feedstock for electrolytic hydrogen production.

Mining, extraction, and recycling technologies of critical metals consume high energy and are environmentally sensitive. Due diligence for the environment is necessary during the processes of critical metals, so that green hydrogen production can be accepted and adopted worldwide.

Recycling of critical metals and REEs is essential to overcome the constraints in the availability of critical metals. Companies should look into supply chain partnerships, such as the opportunities we see in the battery recycling market today, where recyclers are partnering across the battery production value chain. In the electrolyser market, this could mean partnerships among mining players, stack producers, engineering, procurement, construction (EPC) firms, and recyclers. Initiatives have been taken to corroborate and fund sustainable hydrogen manufacturing and electrolyser recycling globally. It has been predicted that by 2035, clients should expect to see complete electrolysers being recycled, as technologies like automated sorting and dismantling, along with metals extraction and purification, become mainstream.

The production of blue hydrogen requires carbon capture, utilisation, and storage (CCUS) technology. Most industrial carbon emissions come from fossil-fuel-run power plants, which account for around 40% of global carbon dioxide emissions and 54% of India's. Though carbon capture technology is well developed, its utilisation by industry to the required level and its storage in an underground aquifer remain uncertain, as they depend on favourable geological and reservoir rock conditions.

India is spearheading the global decarbonisation campaign in terms of investment, implementation, incentivisation, and policy framework. India's NGHM to produce 5 million tonnes of green hydrogen and set a target of 100 million tonnes of coal gasification by 2030 is a testament to its commitment.

The objective of climate policy is solely focused on how much CO_2 reduction can be achieved for each dollar spent. At present, the CO_2 abatement cost through electrolytic green hydrogen is very high, ranging from \$70 to \$175 per tonne of CO_2, because of its high production costs. This makes sense for wealthier nations, which have abundant capital and infrastructure and can absorb the risk of scaling up. As technology develops and matures, and as demand grows, the cost of production reduces and ROI increases.

Electrolyser technology is in a growing stage, and therefore, many more advancements in this field are expected. Solid Oxide and Anion Exchange batteries are emerging technologies that are expected to deliver high efficiency at lower cost, because they do not need costly electrodes and electrolytes. Looking at this, India should focus more on R&D works in electrolytic technology, rather than making huge investments in existing electrolytic technology. Investment in producing electrolytic hydrogen at this stage should be left to developed nations, who have flooded the atmosphere with carbon, in the race of rapid industrialisation from the 1970s onwards, leaving no space for developing countries. India should join the race of electrolytic hydrogen production on a large scale at a later stage, when this technology is mature and becomes more economical. It should focus more on blue hydrogen through coal gasification and other thermogenic and biogenic pathways, while pursuing the electrolytic hydrogen race. Coal gasification pathway to produce blue hydrogen should be encouraged, where carbon capture and the cost of hydrogen production become more economical. Integrated coal gasification combined cycle (IGCC) power plants should replace the direct firing of coal. India holds a high reserve of coal, 9.5% of the global coal reserve.

According to estimates from the International Renewable Energy Association (IRENA) and the EU Commission, the share of clean hydrogen in the global energy mix is likely to be between 12 and 24% by the middle of the century. Projections indicate that by 2050, green hydrogen will make up the largest share, around two-thirds of the global hydrogen supply mix, primarily because of falling renewable and electrolyser costs as per Global Energy Prospective 2023 by McKinsey.

A complete overview of clean hydrogen, taking into account all the above-mentioned aspects, aiming for global decarbonisation, has been discussed.

Limitations to bringing clean hydrogen to the fuel market, along with measures to overcome them and the future landscape of clean hydrogen in the global energy mix, have been covered to the extent possible. Utmost efforts have been made to make the book understandable to readers and make them aware of the role of clean hydrogen in the current energy transition phase.

While writing the manuscript, I have referred literatures issued from prestigious academic institutes, R&D organisations, national and international energy agencies, government institutions, technical and research councils, and production industries. These references have been mentioned at the point of discussion in the relevant chapter. While my previous book "Sustainable Energy and Environment – Achieving Net Zero Emission" talks about the adverse impact of fossil fuel combustion on the environment with ways and means to overcome it, this book "Hydrogen – Energy of the Future" is focussed on how clean hydrogen can be the best alternative to fossil fuel to address the global sustainability issue more effectively.

Target readers of the book may be energy experts, environmentalists, scientists, engineers, policy makers, investors, academia, and those who are interested in knowing about the future of energy and the environment.

Hope the reader will like it. Your feedback, which shall be a great input to enrich me in my quest for learning and sharing, can be mailed to author. eande@gmail.com

With warm regards
SK Pandey

1 Genesis of Hydrogen and Development

Hydrogen is a unique element, created first from the Big Bang explosion, where infinite energy was released. The energy condensed into mass, resulting in the formation of the universe. As per the article issued in 2019 in the *Journal on Astronomy*, hydrogen was formed within 3 minutes after the Big Bang explosion, when space cooled enough to form hydrogen nuclei with one proton. This is called the era of nucleosynthesis. Some of these nuclei combined to form helium as well, though in much smaller quantities. Therefore, hydrogen is sometimes called the three-minute element by astrophysicists.

Astrophysicists from NASA estimate that the universe is comprised mainly of hydrogen and helium. These 2 elements make up 98% of the visible matter in the universe, consisting of about 74% hydrogen and 24% helium. Nevertheless, our world and everything it contains–even life itself–is possible only because of the existence of heavier elements such as carbon, nitrogen, oxygen, silicon, iron, and many, many others.

Today, hydrogen has received much attention as an environmentally friendly option to help decarbonise the planet and thereby overcome global warming. The road to understanding hydrogen's potential offers a fascinating tour through scientific discovery and industrial ingenuity. The history of hydrogen, presented as a timeline from its discovery to present-day use, is given below.

- 1766 – Hydrogen was first identified as a distinct element by British scientist Henry Cavendish after he separated hydrogen gas by reacting zinc metal with hydrochloric acid. In a demonstration to the Royal Society of London, Cavendish applied a spark to hydrogen gas, yielding water. This discovery led him to later find that water (H_20) is made of hydrogen and oxygen.
- 1783 – Jacques Alexander Cesar Charles, a French physicist, launched the first hydrogen balloon flight. Known as "Charlier", the unmanned balloon flew to an altitude of 3 kilometres. Three months later, Charles himself flew the first manned hydrogen balloon.

- 1788 – Building on the discoveries of Cavendish, French chemist Antoine Lavoisier gave hydrogen its name, which was derived from Greek words "hydro" and "genes", meaning "water" and "born off" respectively.
- 1800 – English scientists William Nicholson and Sir Anthony Carlisle discovered that applying an electric current to water produced hydrogen and oxygen gases. This process was later termed "electrolysis".
- 1839 – The fuel cell effect, combining hydrogen and oxygen gases to produce water and an electric current, was discovered by Swiss chemist Christian Friedrich Schoenbein.
- 1845 – English scientist and judge Sir William Grove demonstrated Schoenbein's discovery on a practical scale by creating a "gas battery". For this achievement, he earned the title "Father of the Fuel Cell".
- 1920s – German engineer Rudolf Erren converted the internal combustion engines of trucks, buses, and submarines to run on hydrogen or a hydrogen mixture. British scientist and Marxist writer JBS Haldane introduced the concept of renewable hydrogen in his paper Science and the Future by proposing that "There will be great power stations where, during windy weather, the surplus will be used for the electrolytic decomposition of water into oxygen and hydrogen."
- 1958 – The US formed the National Aeronautics and Space Administration (NASA). NASA's space programme currently uses the most liquid hydrogen worldwide, primarily for rocket propulsion and as a fuel for fuel cells.
- 1959 – Francis D. Bacon of Cambridge University in England built the first practical hydrogen fuel cell of a 5-kilowatt system that powered a welding machine. He named this fuel cell design "Bacon Cell".
- 1970 – Electrochemist John O'Mara Bockris coined the term "Hydrogen Economy". He later published "The Solar-Hydrogen Alternative," describing his vision of a hydrogen economy where cities in the United States could be supplied with solar energy.
- 1973 – The Organisation of Petroleum Exporting Countries (OPEC) oil embargo and the resulting supply shock suggested that the era of cheap petroleum had ended, and that the world needed alternative fuels. The development of hydrogen fuel cells for conventional commercial applications began.
- 1998 – Iceland unveiled a plan to create the first hydrogen economy by 2030.
- 1999 – Europe's first hydrogen fuelling stations were opened in German cities of Hamburg and Munich. The Royal Dutch/Shell company committed to a hydrogen future by forming a hydrogen division.

- 2001 – Ballard Power Systems launched the world's first PEM fuel cell system, designed for the integration into a wide variety of industrial and consumer end-product applications.

Since the start of the twenty-first century, the importance of hydrogen has become increasingly evident for decarbonising the planet, which is necessary to restrict and reverse the rise in global temperatures.

The rise in global warming is primarily due to the combustion of fossil fuels that emit a large amount of carbon dioxide, more than the natural carbon cycle can absorb. It has been estimated that the natural carbon cycle can absorb only around 60% of anthropogenic carbon dioxide emissions, and the remaining 16 billion tonnes per year accumulate in the atmosphere, raising global temperatures. Carbon dioxide present in the atmosphere absorbs the heat from the sun's rays as well as sunlight reflected from the Earth. It works like a blanket on Earth, preventing solar heat from dissipating into open space and raising Earth's temperature.

Hydrogen can be the most suitable alternative to fossil fuels for decarbonising the planet because it emits no carbon dioxide on combustion, but only steam/water vapour, and is abundantly available on earth in compounded form. It can be made available in any part of the world, like natural gas, through packaging and transportation.

Unfortunately, hydrogen is not available in free form in nature, but in compounded form, mostly in the form of water, which is abundantly available in nature. Hydrogen must be separated from water for use, which requires energy and costs. The energy should come from a source that does not emit carbon. The production of hydrogen should be economically viable and affordable to the people at large, and the risks involved in handling should be as low as practicable. The world is finding out the ways and means to overcome these constraints to bring hydrogen into the energy market, which should be techno economically viable, to mitigate the menace of global warming.

Earlier, hydrogen was used in chemical manufacturing as a reducing agent, as a laboratory reagent by the industries, and as a catalyst, but not as an energy source for public utilities. But today the versatility of hydrogen is realised—it can be used as an energy carrier and energy storage, in addition to chemical manufacturing. It is called the Fuel of the Future because it has the potential to be an alternative to fossil fuels and can be made accessible to any corner of the world through pipes and packages, where other forms of energy, like solar, wind, and electricity, may not reach.

Never before has hydrogen been in focus as it is today. Since the start of the twenty-first century, the progress in finding new materials, production technology, storage, and transportation for hydrogen is immense, so much so that market players all over the world, who are experts in assessing the future, have predicted that by 2050, hydrogen will contribute around 15% in the global energy mix.

It is strongly believed that global consciousness of the environment, awareness of global warming and commitment to mitigate it, and progress in science and technology to find ways and means to make green hydrogen production techno economically viable are making it the best alternative to natural gas as the Fuel of the Future.

A vision of hydrogen by Jules Verne in his novel "The Mysterious Island", way back in 1974, is quoted below.

"*I believe that water will one day be employed as fuel, that hydrogen and oxygen, which constitute it, used singly or together, will furnish an inexhaustible source of heat and light of an intensity of which coal is not capable.*"
Jules Verne, *The Mysterious Island (187)*

2 Global Warming and Role of Hydrogen

The world is looking for an alternative to fossil fuels to fight global warming arising due to greenhouse gas emissions. Fossil combustion is the major source of greenhouse gas emissions. More than 85% of global primary energy is supplied from fossil fuels, consisting of coal, oil, and natural gas, which emit the most greenhouse gases. Carbon dioxide (CO_2) is the largest of all greenhouse gases in anthropogenic emissions, which alone shares 74–75% of global emissions. Other greenhouse gases are methane, nitrous oxide, chlorofluorocarbons, and sulphur hexafluoride.

In addition to the combustion of fossil fuels, greenhouse gases come from agriculture, industrial byproducts (mainly cement and refineries), buildings, forestry, waste, and other sectors. Greenhouse gases other than CO_2, such as methane, nitrous oxide, and sulphur hexafluoride, are produced during the combustion of fossil fuels, though in very small quantities—in fractions of a percent or ppm—but these gases have much higher global warming potentials than CO_2. These gases contribute approximately 22–25% to the total global warming impact. Reducing fossil combustion will not only reduce CO_2 but also other greenhouse gases, which are produced simultaneously during combustion.

2.1 Greenhouse Gas Emission and Global Temperature Rise.

Greenhouse gases are gaseous molecules in the atmosphere that absorb infrared radiation (heat) reflected from the Earth's surface, preventing it from escaping to space. This process is known as the greenhouse effect. It is a natural phenomenon that keeps the Earth warm enough to support life. But a rise in Earth's temperature beyond a certain limit has an adverse impact on the ecological balance, threatening the survival of life on Earth. Human activities, particularly the combustion of fossil fuels and deforestation, have significantly increased the concentration of greenhouse gases in the atmosphere, leading to a rise in global temperatures. Forests and other green cover act like a sink for CO_2, which helps in maintaining the Earth's temperature. Increased greenhouse gas concentrations can lead to a range of climate impacts, including rising sea levels, extreme weather events, and shifts in the ecological balance—a balance between atmosphere, hydrosphere, and biosphere. It has adverse effects on life on Earth.

There are natural and human-caused greenhouse gases. Natural sources include plant and soil respiration, decomposition of plants and animals, and ocean respiration. Greenhouse gas emissions, other than carbon dioxide, occur naturally as well as from human activities, such as water vapour, methane, and nitrous oxide. Some greenhouse gases come from human-made synthetics, such as chlorofluorocarbons (CFCs), hydrofluorocarbons (HFCs), perfluorocarbons (PFCs), and sulphur hexafluoride (SF_6). They can be found, for example, in aerosol sprays, air conditioning, refrigerants, and electronics.

Human-caused emissions come from burning fossil fuels such as oil, coal, and natural gas, as well as from activities like deforestation, agriculture, and cement production. According to the Intergovernmental Panel on Climate Change (IPCC), human activities are responsible for almost all of the increase in greenhouse gases in the atmosphere over the last 150 years.

Integrated Carbon Observation System (ICOS) measures the most significant greenhouse gases in the atmosphere, ecosystems, and ocean: carbon dioxide (CO_2), methane (CH_4), and nitrous oxide (N_2O).

Greenhouse gas (GHG) emissions originate from burning fossil fuels for energy, industrial processes, and certain agricultural practices. Electricity production, transportation, steel, and cement industries are major contributors. Agriculture, commercial/residential activities, deforestation, and land use contribute significantly.

Greenhouse gas emissions, broken down by sector, are shown in Figure 1. It can be seen from the figure that 73.2% of GHGs are emanating from the energy sector, followed by agriculture, forestry, and the land-use sector. The land-use sector takes into account the emissions caused by urbanisation, loss of habitats, and deforestation.

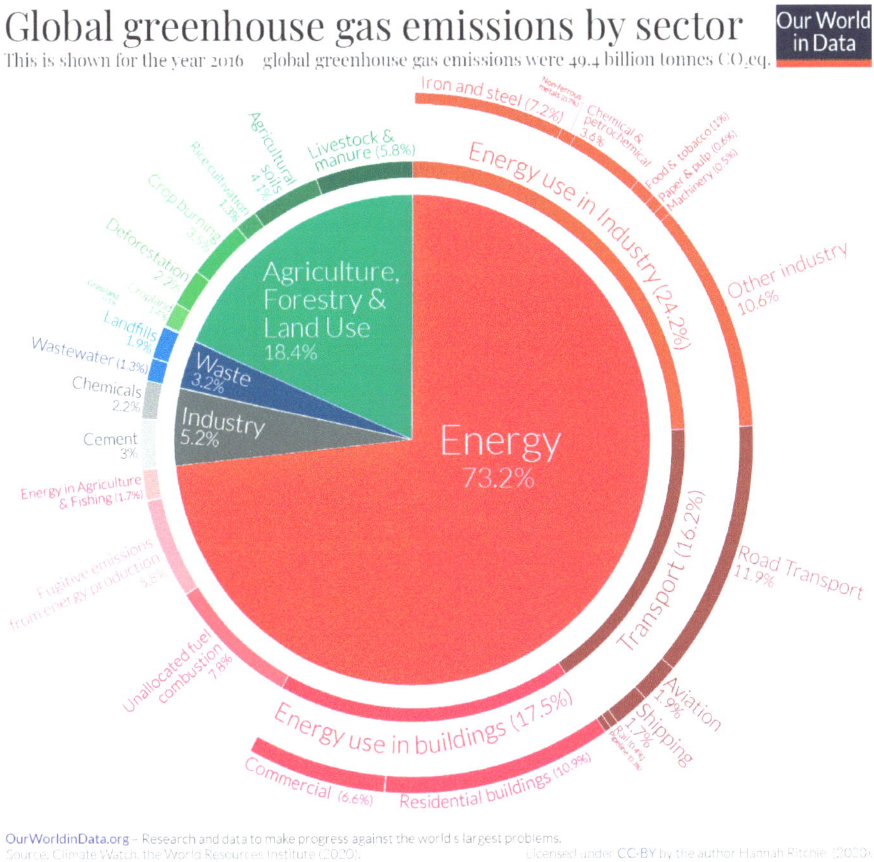

▲ Figure 1: Global Greenhouse Gas Emission by Sector

There is a scientifically established relationship between atmospheric CO_2 concentrations and rising global temperatures. This was established after 40 years of consistent observation and monitoring by Charles David Killing at the Scripps Institution of Oceanography, at a strategic location on Mauna Loa, Hawaii. The Mauna Loa Observatory (MLO) in Hawaii is a critical strategic location for monitoring global temperature rise because it provides the world's longest, most continuous record of atmospheric carbon dioxide concentrations. Since CO_2 is the primary greenhouse gas, driving global warming, the data collected there—known as the Keeling Curve—serves as the foundational, uncontaminated baseline for global climate science. Since the annual change in Earth's absolute temperature is extremely low, it is monitored as a temperature anomaly.

A temperature anomaly is the difference between an observed temperature and a reference or baseline temperature, usually an average over a long period. Positive anomalies indicate temperatures warmer than the baseline, while negative anomalies indicate temperatures cooler than the baseline. These anomalies are crucial in climate science because they help track changes over time and minimise the impact of factors like location or station differences.

The trend in Earth's temperature since the pre-industrial period is shown in Figure 2.

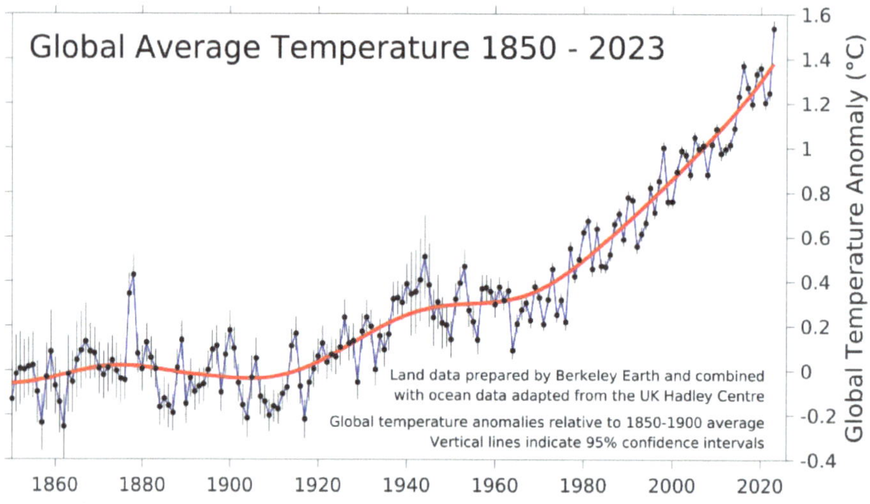

▲ Figure 2: Global Average Temperature 1850-2023

2.2 Fossil Fuel and Global Warming

Global warming can be restricted and reversed if greenhouse gases, which are emitted mainly from the combustion of fossil fuels, are reduced. However, it is difficult to do so because it is the primary source of energy for power generation, industrial manufacturing, and domestic consumption all over the world. Fossil fuels account for the largest portion of the global energy mix, around 82% of the world's primary energy. This share has remained relatively stable over the past few decades, despite increasing deployment of renewable energy sources. Today, around one-third of global energy generation in the form of electricity comes from coal, which produces the highest carbon dioxide emissions on combustion when compared with gas and oil. Primary global energy consumption from different sources, as reported in the *Statistical Review of World Energy-24*, is shown in Figure 3.

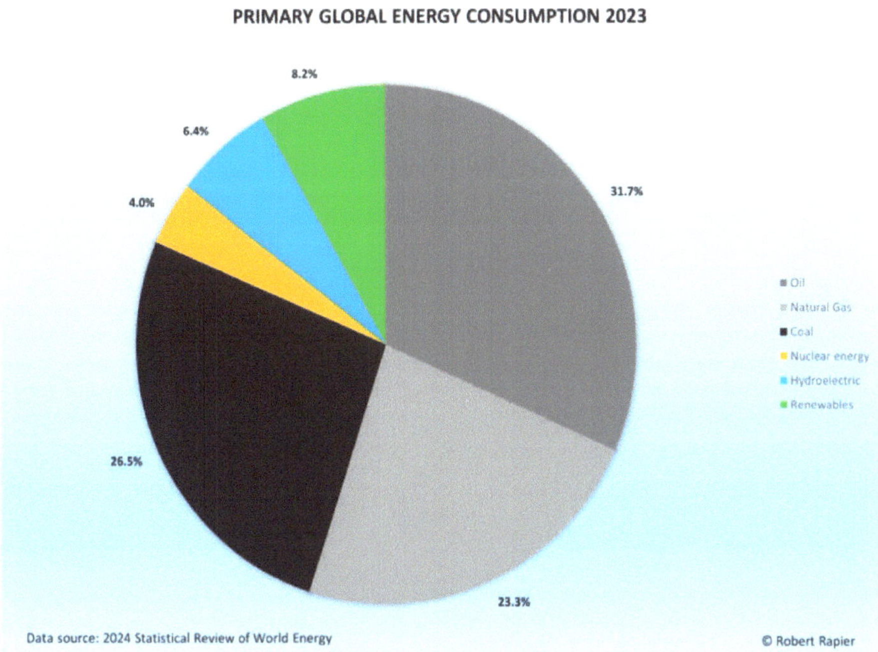

▲ Figure 3: Primary Global Energy Consumption 2023

In 2024, global carbon dioxide emissions from fossil fuels reached a record high of 37.4 billion metric tonnes, representing a 0.8% increase compared to 2023, according to the Global Carbon Budget. This increase puts the world further away from meeting climate targets and indicates that global emissions have not yet peaked. In addition, there are other sources of CO_2 emissions from industrial sectors, mainly the steel and cement industries. Total CO_2 emissions from fossil fuels, land use, and other sources are projected to set a new record at 41.6 Gt CO_2 (gigatonnes of carbon dioxide equivalent) in 2024, reflecting a growth of 2% from 2023 levels. Total greenhouse gas emissions, including other high-global-warming-potential chemicals like methane, nitrous oxide, HFCs, and SF6, amount to around 51 billion tonnes of CO_2 equivalent when accounted for their global warming impacts.

Figure 4(a) and 4(b) indicate CO_2 emission from fuel or industry type, and world greenhouse gas emission region-wise, respectively.

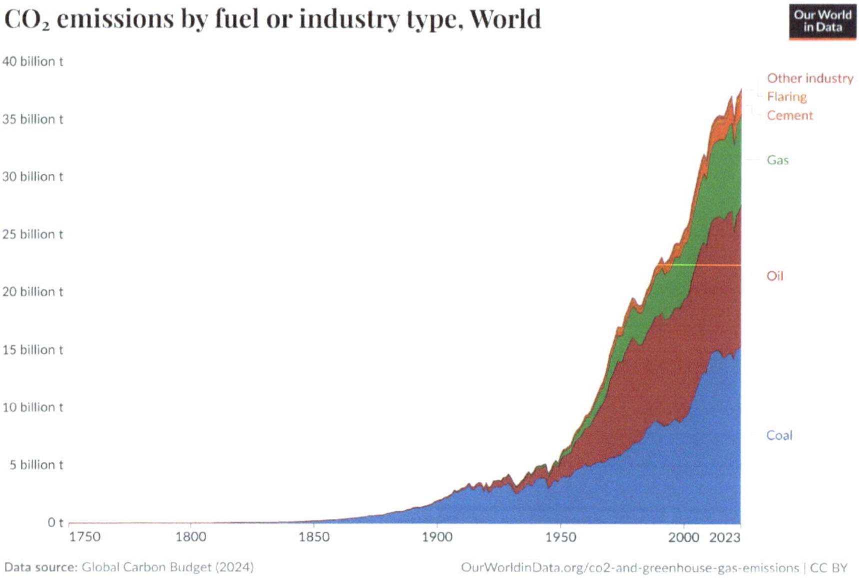

▲ Figure 4(a): Global CO2 emission by fuel and industry type

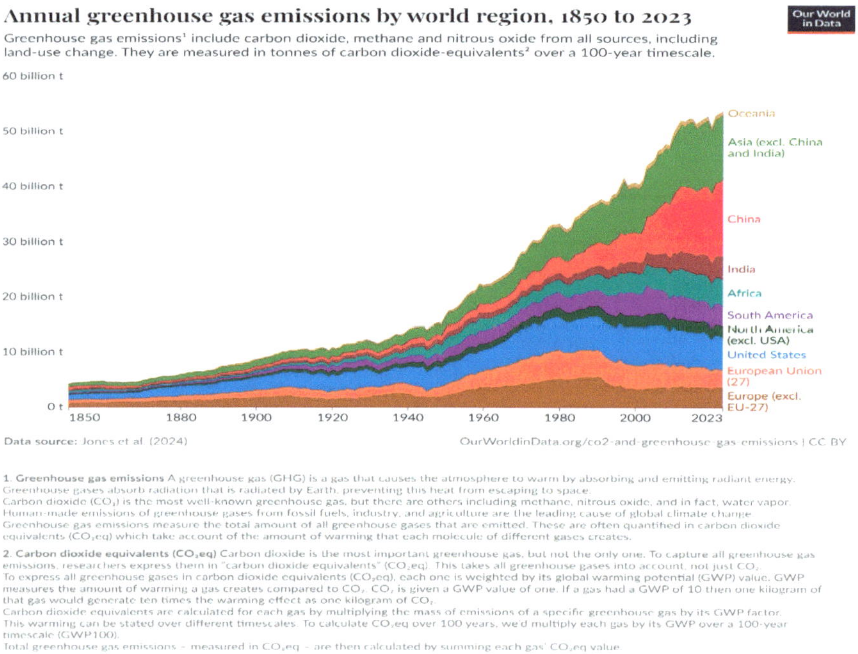

▲ Figure 4(b): Global Annual Greenhouse Gas Emission 1850-2023

It is worth mentioning here that steam or water vapour is a potent greenhouse gas because it absorbs infrared radiation or heat. But it releases the absorbed heat within a few days, or even hours, as latent heat of condensation, which is dissipated into space. Human activities don't release a large amount of water vapour into the atmosphere, which can lead to increased temperature. It amounts to nothing compared with oceanic evaporation. Water vapour is the most abundant greenhouse gas in the atmosphere and plays a crucial role in the natural greenhouse effect that keeps the Earth warm enough, contributing to about 60% of the overall effect.

Unlike other greenhouse gases, which stay in the atmosphere for a long time (100 years), water vapour has a much shorter residence time, typically only hours or days. It is unique because it can easily condense from a gaseous state to a liquid, which is a key part of the water cycle.

2.3 Global Warming Potential

Global Warming Potential (GWP) is a metric that compares the warming effect of different greenhouse gases to that of carbon dioxide (CO_2). It quantifies how many times more heat a greenhouse gas can trap in the atmosphere over a specified period (usually 100 years) relative to the same mass of CO_2 (usually 1 tonne). Gases with higher GWPs have a greater warming impact than CO_2 for the same time period and mass. GWP is a relative measure of warming impact, relative to CO_2, which is taken as having a GWP of 1.

The United Nations Framework Convention on Climate Change (UNFCCC), held in Kyoto, Japan, in 1997, agreed to reduce the emission of greenhouse gases, which are harmful to human life. The list of greenhouse gases covered by the Kyoto Protocol, their GWP values, and atmospheric lifetimes is given in Table 1. This came into force in 2005.

The protocol established legally binding emission-reduction targets for industrialised countries, with the first commitment period starting from 2008 to 2012. While the protocol was a significant step, it faced criticism for its limited scope and for the wit*hd*rawal of some key nations, including the United States. The Kyoto Protocol could not serve the purpose as expected. The Paris Agreement on climate change, adopted by the UNFCCC in 2015, is more effective than the Kyoto Protocol.

▼ Table 1: Global Warming Potential for Major Greenhouse Gases

Greenhouse gas	Chemical formula	Radiative Efficiency (W/m^2ppb)	Atmospheric Lifetime (years)	GWP (100-year time horizon)
Carbon dioxide	CO_2	0.0000133	multiple*	1
Methane	CH_4	0.000388	12	28
Nitrous oxide	N_2O	0.0032	109	273
Select halogenated compounds:				
Chlorofluorocarbon-12 (CFC-12)	CCl_2F_2	0.358	102	12,500
Hydrofluorocarbon-23 (HFC-23)	CHF_3	0.191	228	14,600
Nitrogen trifluoride	NF_3	0.204	569	17,400
Sulfur hexafluoride	SF_6	0.567	1,000	24,300

NOTES

*Carbon dioxide is quite stable in Earth's atmosphere, but individual carbon dioxide molecules are in near-constant flux to and from different reservoirs, such as the surface ocean, land plants and animals, soils, and the lithosphere. Some of these fluxes operate on the scale of years, and others on the scale of millennia. As such, no single lifetime can be given for carbon dioxide because it moves throughout the earth's system at differing rates. A convenient and commonly used estimate for the average CO2 atmospheric lifetime is 100 years.

SOURCE

Sixth Assessment Report (Intergovernmental Panel on Climate Change, 2021).

2.4 Fossil and Hydrogen Fuels

Hydrogen, on combustion, does not produce CO_2 or any other GHG. It produces only steam. This is the biggest advantage of hydrogen, which is why it is preferred as the best alternative to fossil fuels. Secondly, hydrogen has a heating value 3 times that of natural gas per unit mass, in terms of higher heating value (HHV). This means one-third of hydrogen can replace methane in terms of mass. However, in terms of volume, hydrogen carries much less heat, one-third of methane. Hydrogen is abundantly available on Earth in compound form.

A comparative statement of properties of hydrogen with other fuels is given in Table 2.

▼ Table 2: Properties of H_2 gas and other fuel type, comparison

Fuel/property	Hydrogen	Natural Gas	Petrol	LPG
Lower heating value (MJ/kg)	120.7	49.54	41.87 - 44.19	46.05
Higher heating value (MJ/kg)	141.89	54.89	43.73 - 47.45	50.24
Density at standard conditions (kg/m^3)	0.08	0.6	720 – 780	510
Phase at standard conditions	Gas	Gas	Liquid	Liquid
Auto-ignition temperature[1] in air (°C)	566 – 582	540	257	454 – 510
Ignition limit[2] in air (Vol%)	4.1 – 74	5.3 – 15	1.4 – 7.6	2.2 – 9.5
Diffusion coefficient[3] in air (cm^2/s)	0.61	0.16	0.05	0.11

1. *Auto-ignition temperature is the lowest temperature at which a fuel will ignite when an external source of ignition is present*
2. *Ignition limit is the range of concentration within which the fuel will ignite, if an ignition source is present*
3. *Diffusion coefficient is used to determine the rate at which the fuel disperses (the higher the co-efficient, the faster the rate)*

Replacing natural gas with hydrogen shall avoid the emission of GHG gases which, in turn, shall reduce global warming. Sourcing H_2 from fossil fuels will not serve the purpose of avoiding emissions, because this process will co-produce carbon dioxide and other toxic gases which, finally, shall be emitted into the atmosphere. Therefore, hydrogen must be produced from a non-fossil source, such as solar, wind, biomass, geothermal, or nuclear. Energy produced from non-fossil resources, where there are no CO_2 emissions, is called green hydrogen. Biomass is considered a source of green energy because the CO_2 produced by its combustion is fully recycled through photosynthesis, without any CO_2 accumulation in the atmosphere. It is assumed that enough green plantation would be done that would absorb the equivalent amount of CO_2 produced by biomass combustion. Plants and trees absorb CO_2 through photosynthesis, which takes place in the presence of sunlight. Photosynthesis is the process by which CO_2 is absorbed by moisture (water) in the plant, in the presence of sunlight, producing carbohydrates and oxygen, and chlorophyll present in the plant catalyses the reaction. Plants use carbohydrates to grow and release oxygen into the atmosphere. The photosynthesis reaction is

$$6CO_2 + 6H_2O = C_6H_{12}O_6 + 6O_2$$

Hydrogen is the most abundant element available in the universe, in compounded form. Most of this is bound up in the form of water, which covers two-thirds of the globe's surface. Sixty percent of the atoms in our bodies are hydrogen. It is present alongside carbon in most organic compounds, such as carbohydrates in our food, made from hydrogen, carbon, and oxygen, which is why they are called carbohydrates.

Oil and gas fuels are made from hydrogen and carbon in varying proportions and are called hydrocarbons. When a hydrocarbon combines with oxygen in a combustion reaction, it produces CO_2, water, and heat energy; this reaction is called an exothermic reaction. Such a reaction can be represented as:

$$CH_4[gas] + 2O_2[gas] \rightarrow CO_2[gas] + 2H_2O[steam] + 891 kJ \text{ heat released.}$$

Oil and gas contain carbon and hydrogen. Coal contains oxygen additionally, due to its formation in an environment different from that of oil and gas. Oxygen in coal exists in both organic and inorganic forms. Organic oxygen is primarily found in the form of oxidised organic compounds within the coal matrix. Inorganic oxygen is associated with mineral matter, such as oxides, hydroxides, and sulphates, and influences chemical and physical properties, as well as behaviour during combustion and other processes. Hydrogen and carbon contribute to the heating value, but the presence of oxygen reduces it because it occupies the space in unit mass, sharing no heat value. The carbon content, hydrogen content, and carbon dioxide emissions from different fossil fuels during combustion are shown in Table 3.

▼ Table 3: CO_2 emission from different fuel

	Carbon Content	Hydrogen Content	Heating Value	CO_2 emission in kg per MWh energy produced (kg)
Coal	more than 90%	5%	17.4 to 23.9 MJ/kg	900
Crude Oil	84-87%	11-13%	42 to 47 MJ/kg	565
Natural Gas	75%	25%	42 to 55 MJ/kg	365
Hydrogen	0%	100%	120.7 MJ/Kg (LHV)	0

2.5 Classification of Hydrogen

Hydrogen is divided into 3 categories based on CO_2 emissions during manufacturing, each represented by a colour. These are grey hydrogen, blue hydrogen, and green hydrogen.

Grey hydrogen is the one that is produced from fossil fuels (mostly methane gas). Carbon dioxide produced during the process is released into the atmosphere. Today, more than 95% of hydrogen produced globally across all industries is grey hydrogen from fossil fuels. Since the stoichiometry of the SMR reaction involves converting methane (CH_4) to hydrogen and carbon monoxide (CO), and then CO to CO_2 through the Water Gas Shift Reaction, all the carbon in methane ends up forming CO_2.

Greenhouse gas (GHG) emissions in hydrogen production from natural gas, primarily through the SMR process, range from 7 to 12 kg CO_2-equivalents per kg of hydrogen produced in terms of well-to-gate emission. This range accounts for variations in natural gas extraction, transportation, and processing efficiencies, as well as the specific SMR process. It's important to note that these figures are not just for CO_2 emissions but also include other greenhouse gases such as methane emitted throughout the life cycle, expressed in CO_2 equivalent. However, the majority of the emissions are CO_2.

Blue hydrogen is also produced from fossil fuels, mostly through the SMR process, but in this case, the carbon dioxide produced during production is captured and either utilised or stored. Blue hydrogen is produced from biomass and fossil fuels by using CCUS technology, which makes the process carbon negative. Blue hydrogen is also considered clean hydrogen. The maximum greenhouse gas emission limit for blue hydrogen production in India is set to be 4.0 kg CO_2e/kg of hydrogen, based on a well-to-gate analysis. This includes emissions from the production facility of oil and gas wells.

Very few industries produce blue hydrogen today. Setting up a carbon capture facility at a power plant incurs an additional capital expenditure of around 30–35% and consumes 25% of the plant's internal power. This increases the levelised cost of clean power.

Green hydrogen does not emit carbon dioxide during production. Hydrogen produced by electrolysis using renewable energy is green hydrogen. In India, the Green Hydrogen Standard defines having a well-to-gate GHG emission of not more than 2.0 kg CO_2 equivalent per kg H_2, the same as the standard set by the Government of Energy (GOE), US. This standard outlines the emission thresholds for hydrogen production, which can be classified as green, produced using renewable energy sources.

Less than 1% of the hydrogen produced all over the world is green. This is because the production cost of green hydrogen is more than 3 times that of grey hydrogen due to high power and electrolyser costs.

Hydrogen produced from biomass is considered green hydrogen because the carbon dioxide emitted is considered to be fully recycled (no accumulation) through the photosynthesis process in new plantation growth. Therefore, if

CO_2 is captured during the production of hydrogen from biomass, the process becomes carbon negative.

In India, grey hydrogen, produced through the steam methane reforming process, costs Rs. 160–200 per kg when produced from natural gas. This costs \$10–13 per MMBtu (Metric Million British Thermal Unit), as against green hydrogen, Rs. 350–450 per kg, produced using solar and wind energy, costing Rs. 2.5–3.5 per kwh. Many technological advancements are being made in electrolysis and renewable power production technology to reduce the cost of green hydrogen so that it can compete with grey hydrogen. It is estimated that by 2030, the cost shall be comparable to that of grey hydrogen.

Figure 5, taken from the World Economic Forum, indicates the range of greenhouse gas emissions for grey, blue, and green hydrogen production.

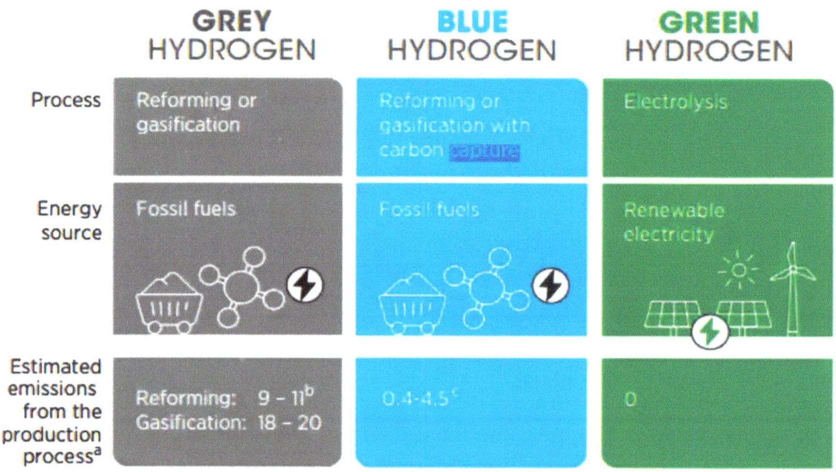

Note: a) $CO_{2\text{-eq}}$/kg = carbon dioxide equivalent per kilogramme; b) For grey hydrogen, 2 kg $CO_{2\text{-eq}}$/kg assumed for methane leakage from the steam methane reforming process. c) Emissions for blue hydrogen assume a range of 98% and 68% carbon capture rate and 0.2% and 1.5% of methane leakage.

▲ Figure 5: Selected colour-code topology of hydrogen production

3 Hydrogen as an Alternative to Fossil Fuels

Green hydrogen can be the best substitute for fossil fuels because it produces only steam on combustion, not carbon dioxide. It is abundantly available in compounded form in water. It can be made available to any part of geography by packaging and transportation, similar to natural gas and LPG. The constraints on bringing hydrogen into energy markets are its high production cost and the risks involved in handling.

As per the IRENA, green hydrogen produced through electrolysis—splitting water into hydrogen and oxygen using electricity—currently costs around $7 per kg, assuming an average solar power cost of $85/MWh and an efficient PEM electrolyser cost $840/Kw, whereas the production cost of grey H_2, through the SMR process, mostly used all over the world, comes out to be $2.2/kg, at the natural gas cost of $8 per MMBtu energy. This means electrolytic H2 is more than 3 times as expensive as conventional H2 at current energy costs.

Handling of hydrogen, like storage, transportation, and dispensation, is challenging due to material and social risks involved in its handling. Hydrogen spreads 3.8 times faster than natural gas and is more likely to catch fire due to its wider flammability range, which is 8 times wider than that of natural gas. It can cause stress cracking in natural gas pipelines, a phenomenon called hydrogen embrittlement; therefore, a material different from what we use for natural gas will be required to handle it in pure form.

The major constraint why hydrogen is not able to compete with natural gas can be explained mainly in the following ways.

1. High Production cost.
2. Difficulties in handling.
3. High societal and material risk.

High Production cost

The cost of green hydrogen is high mainly due to the high costs of renewable power and electrolysers. Today, the cost of green hydrogen varies between

Rs. 244 and 357 per kg, depending upon the renewable energy cost—solar, wind, hybrid, or hybrid RTC. Here, hybrid (RTC) means round-the-clock power supply from both solar and wind with a storage facility (batteries).

Recently, Teri School of Advanced Studies conducted a study on how green energy from various renewable power sources influences the levelised cost of green hydrogen production. It has been concluded that, to bring down the cost of hydrogen to $1/kg so it can compete with natural gas, the cost of renewable energy should be Rs. 1/kWh and the electrolyser should be Rs. 3,000/kW, with a thermal efficiency of 90%.

The present efficiency of the most energy-efficient PEM electrolyser on HHV is reported to be around 70%. The cost of a PEM electrolyser per megawatt (MW) varies but generally ranges from $2.0 million to $2.5 million at present. This cost includes equipment, gas treatment, plant balancing, engineering, procurement, construction, and contingencies, according to Hydrogen Insight (Department of Energy - DOE, US report, May 24).

A comparison of the levelised cost of hydrogen (LCOH) from various renewable sources is given in Figure 6.

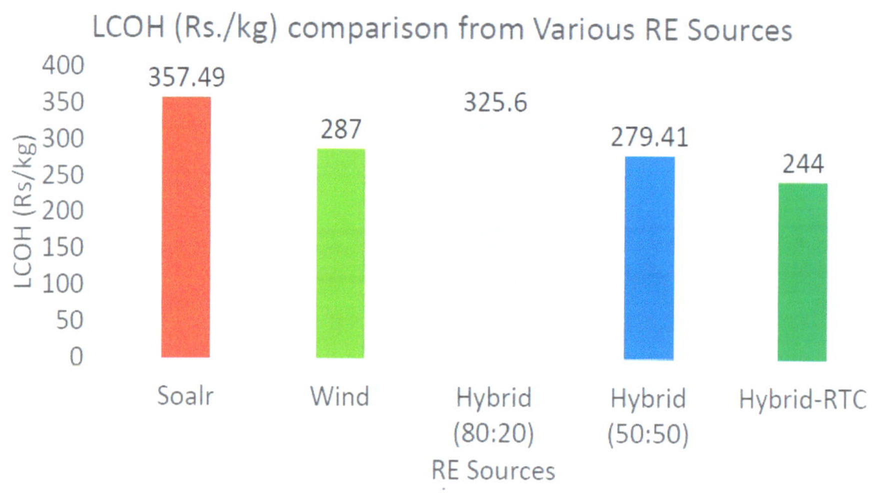

▲ Figure 6: Renewable energy tariffs considered in Rs/kWh are: for solar 2.2, wind 3.4, hybrid (80:20) 2.49, hybrid (50:50) 2.57, and hybrid RTC (SECI) 2.9

3.1 Handling of Hydrogen.

As explained earlier, the storage, transportation, and dispensing of hydrogen are difficult for 2 main reasons: its high diffusivity and high flammability. A comparison of these properties of hydrogen vis-à-vis other fuels is given in Table 2 earlier.

Because of its high diffusivity, hydrogen spreads in the atmosphere much faster, and being the tiniest molecule, it escapes easily through valves and other blockades, causing leakages. Atomic hydrogen penetrates metals and accumulates at the faults of the metal lattice structure, which can cause stress cracking, called hydrogen embrittlement. Therefore, handling hydrogen requires a different material composition than that for natural gas, so that it can sustain hydrogen embrittlement, which adds cost.

The range of explosivity of hydrogen in an air mixture is much larger than that of natural gas, more than 8 times. This makes it more susceptible to explosion. Extra precaution and safety interlocks are needed while handling. A hydrogen flame is nearly invisible, increasing the risk of accidental human contact. The higher flame speed of H2, compared with natural gas, increases the risk of backfire.

The adiabatic flame temperature of hydrogen is higher than that of natural gas. Therefore, nitrous oxide emissions from hydrogen combustion can increase because nitrogen is more prone to form nitrous oxide at higher temperatures. This high flame temperature needs to be controlled to reduce nitrogen oxide formation.

An important concern is the effect of hydrogen on materials. As we know, appliances and infrastructure for handling natural gas are made of a variety of metals and polymers. If hydrogen is handled in these existing appliances and infrastructure, it can affect the properties of metals in different ways. Many appliances use polymer to seal their valves and connections. Sealing materials are typically elastomeric or semi-crystalline thermoplastic. Since hydrogen is the smallest and lightest molecule, much smaller than a methane molecule, it can easily penetrate polymers and change their permeability and stability.

All the above-mentioned handling complications make hydrogen a difficult option to use as a natural gas, requiring changes in the system.

3.2 Blending of Hydrogen with Natural Gas.

Blending hydrogen with natural gas is not new; it can be traced back to the gas produced from coal. This producer gas, a mixture comprising of 30%–50% hydrogen and methane, was being piped during the gaslight era to streetlamps in the early and mid-1800s.

The rapid growth in installed offshore wind and solar power capacity, the reduction in the production costs of renewable power, and the increase in electrolyser capacity reduce the cost of green hydrogen production. Advances in electrolytic technology increase energy efficiency. All these are accelerating the manufacture of green hydrogen production. Recent disruptions to natural gas supply from Russia to Europe, due to changing global geopolitical conditions, have further accelerated the trend towards green hydrogen production in Europe. Blending hydrogen with natural gas in existing natural gas networks is practiced globally, particularly in Europe, which resolves the issue of hydrogen transport to a great extent. The EU aimed to increase the share of hydrogen in Europe's energy mix from its current level of less than 2% to 13–14% by 2050.

The permissible percentage of hydrogen in natural gas networks worldwide varies by region and ranges from 0% to 20%.

UL Solutions has issued a white paper on hydrogen blending in natural gas pipelines that can be transported through the existing infrastructure safely. As per the white paper issued, Global Technical Institute (GTI) viewed that up to 20% hydrogen can be blended with natural gas in pipelines without a significant increase in societal risk. NaturalHy (a project focused on defining the conditions under which hydrogen could be added to natural gas in the existing natural gas system) proposed a limit of up to 50% hydrogen blending in natural gas pipelines without posing risks to pipeline materials and stated that it is safely manageable. However, fire and explosion-related societal risk concerns provide an upper limit of 20% hydrogen blending that can be safely added to natural gas infrastructure. According to the Public Utilities Commission of California, hydrogen-natural gas blends with hydrogen content greater than 20% pose a higher risk of permeating plastic pipes, increasing the risk of ignition of leaked gas. Societal risk limit controls, and a maximum of 20% blending is safely practiced in India and many other nations.

Consolidated information about the challenges and risks involved in using the existing natural gas system for pure hydrogen service is given in Figure 7. The issues, impacts, and challenges involved in transporting pure hydrogen through existing natural gas pipelines are shown in Figure 8. This consolidated information has been collected from LinkedIn, posted by process development experts.

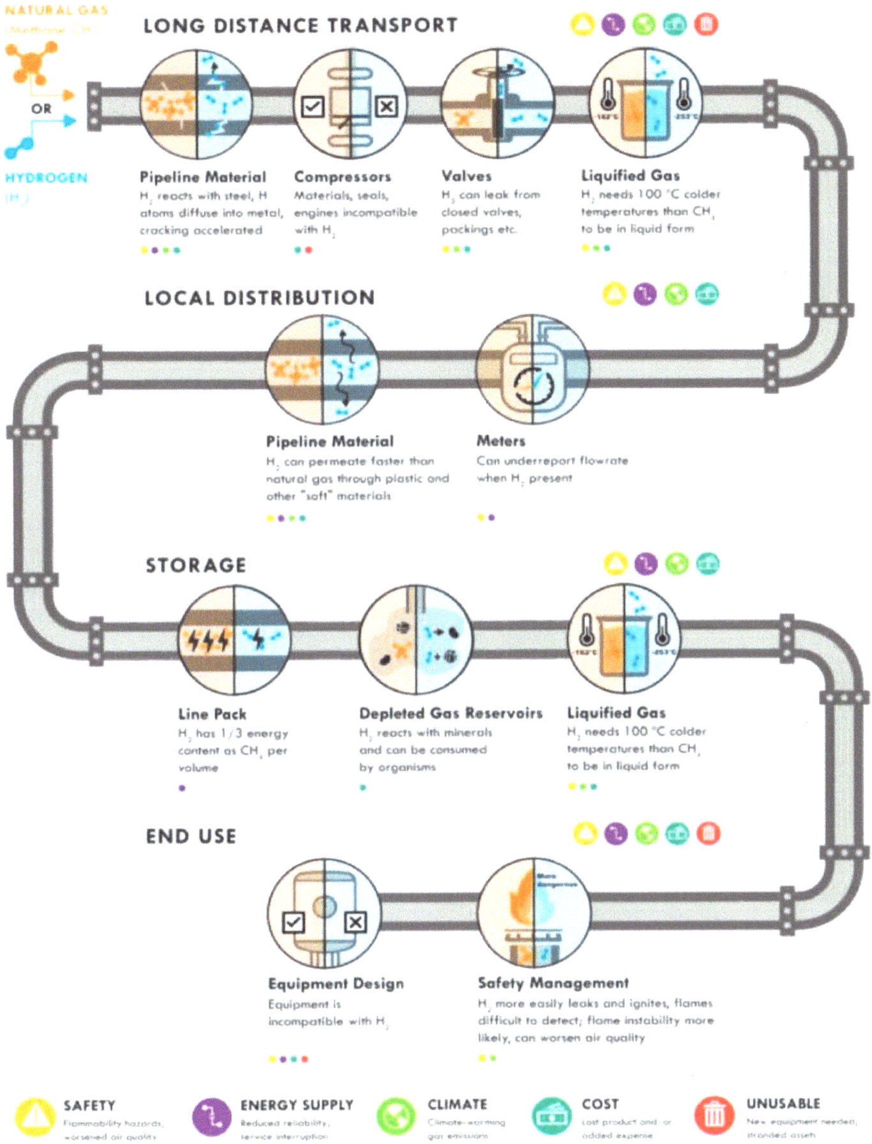

▲ Figure 7: Challenges and risks with using the existing natural gas system for pure hydrogen service

LONG-DISTANCE TRANSPORT

COMPONENT	ISSUES	IMPACTS & RISKS	SOLUTIONS & CHALLENGES
Pipeline material	H_2 can dissociate into atomic H on steel surfaces from corrosion mechanisms, electroplating, or the decomposition of water from damp flux materials during welding	Reduced ductility, fracture toughness and fatigue resistance lead to accelerated cracking, causing leakage and potential mechanical failure [Safety] [Energy supply] [Cost] [Climate]	1. Limit amount of H_2 but limited decarbonization potential [Climate] 2. Add liner/coating but more research needed; difficult to access/ inspect because most pipes underground; leakage; service interruption; expensive [Energy supply] [Cost] 3. Reduce operating pressure but reduces energy supply, line pack unless new pipe added in parallel [Energy supply] [Cost] 4. Replace pipe but service interruption; expensive [Unusable]
Line capacity	H_2 has 1/3 energy content as natural gas per volume; if design pressure de-rated, further reduction in energy content	At least 3x less energy delivered to consumers [Energy supply]	1. Limit amount of H_2 but limited decarbonization potential [Climate] 2. Increase volumetric flowrate by 3x but then compressors require 3x energy [Cost] [Climate]
Compressors	Materials of construction, seals, engines, not designed for H_2	Incompatible equipment [Unusable]	1. Limit amount of H_2 but limited decarbonization potential [Climate] 2. Replace compressors but inconvenient, expensive [Cost] [Unusable]
	Changes in operational pressure or flowrate to resolve other issues requires more energy	More input energy required to compress H_2 [Cost] [Climate]	1. More compressors but inconvenient, expensive [Cost] 2. Shorter distances between compressors but inconvenient [Energy supply] 3. More energy input but takes energy away from other uses [Climate]
Valves	Not designed for H_2	H_2 can leak from closed valves, packings etc. [Safety] [Climate] [Cost]	Replacements required [Cost] [Unusable]
Liquified gas	Requires extremely low temperatures	Energy intensive and leakage [Safety] [Cost] [Climate]	Specialized infrastructure is required [Cost] [Unusable]

SAFETY	ENERGY SUPPLY	CLIMATE	COST	UNUSABLE
Flammability hazards; worsened air quality	Reduced reliability; service interruption	Climate-warming gas emissions	Lost product; and/ or added expense	New equipment needed; stranded assets

▲ Figure 8: Existing natural gas transmission system suitability for pure hydrogen

4 Hydrogen Production

Hydrogen is present in abundance on Earth in compound form, mostly in water. Two-thirds of Earth's surface is covered with oceanic water. The real challenge is separating hydrogen from water, which requires more energy than one can get from. At the present stage of technical development, only around 60–65% of energy efficiency, on the basis of the lower heating value of hydrogen, can be obtained in practice in an electrolytic process of H_2 separation from water.

Since hydrogen is not available in nature directly, like natural gas, and must be extracted from other primary sources, it is not called a primary energy source, but an energy carrier.

Hydrogen production by any process, e.g., electrolysis, reformation, thermogenic, or biogenic, is a process of energy transformation. The process of hydrogen production is always associated with energy losses. When electrical energy is converted into chemical energy in the form of hydrogen through electrolysis, 30–35% of the energy is lost.

Hydrogen is produced through the following sources:

1. Fossil Fuels - coal, oil, and gas.
2. Biomass.
3. Water Electrolysis.

It is worth mentioning here that hydrogen is reported to be found naturally beneath the Earth's surface. A growing number of countries are actively engaged in exploring and extracting natural hydrogen, including Australia, France, Mali, the United States, and Canada. It is thought that natural hydrogen is formed by the natural reformation of natural gas and water under favourable geochemical conditions, occurring beneath the Earth.

As reported recently (Economic Times, 16 March 2024), a reserve of 46 million tonnes of hydrogen has been found underground in Europe, in the Pyrenees and Alps mountain range, but it has yet to be demonstrated.

Green hydrogen can be produced from solar, wind, and nuclear power sources more efficiently. Sourcing hydrogen from biomass is a slow process and not as efficient as the electrolytic process because of the biological and thermochemical processes involved in it, in addition to being logistically cumbersome. The biological process is controlled by bacteria, which are slow. The thermochemical process involves heating and pressure, which result in process heat losses. The electrolytic process is preferred due to its compactness and efficiency.

Energy efficiency in biomass thermogenic process ranges between 40–70%, as against electrolysis, 60–70% through PEM electrolysers. Biogenic process report 20–30% system efficiency.

It is reported that, to produce 100 kg of hydrogen from biomass, through a thermogenic process, 1 ton of dry biomass is required. Technical Readiness Level (TRL) for electrolytic process is rated highest, 8–9, against biomass thermogenic 5–7. Biogenic is rated 4–6 on the TRL scale.

A sketch of pathways through which hydrogen can be produced is given in Figure 9.

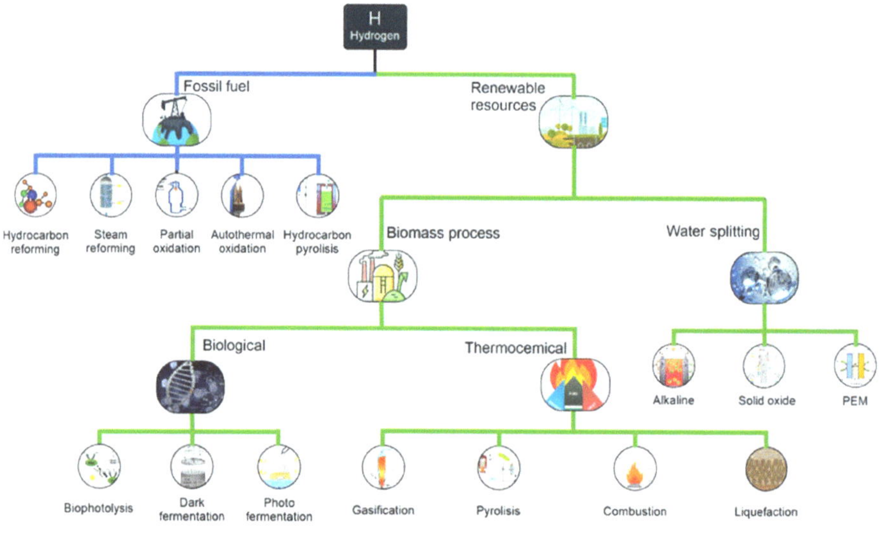

▲ Figure 9: Pathways for Hydrogen Production

More than 98% of hydrogen today is produced from fossil fuels. Oil and gas are relatively much richer in hydrogen than coal, which contains many inorganic impurities. According to the International Energy Agency (IEA) Future of Hydrogen report, June 2019, more than 78% of global hydrogen is produced from fossil fuels—59% from natural gas and 19% from coal—with essentially zero carbon capture. In addition, hydrogen is produced from oil during the reforming and cracking processes in refineries as a byproduct, accounting for around 22% (per IRENA) of total production from fossil fuels.

As of today, very little hydrogen is produced through the electrolysis process globally, reportedly less than 1%.

4.1 Production of Hydrogen from Coal

Hydrogen production from coal using carbon capture technology can provide a low-cost, low-emission, high-volume stream of hydrogen for clean energy.

The Great Plains Synfuel Plant in North Dakota, USA, uses brown coal gasification with CCS to produce hydrogen. Another example is the Hydrogen Energy Supply Chain (HESC) Project in Victoria, Australia, which has demonstrated the extraction and shipment of liquid hydrogen from a coal and biomass mix.

Hydrogen from coal is produced through the gasification route. Gasification is the partial oxidation of coal with steam and a limited oxygen supply to convert it to hydrogen and carbon monoxide (CO). CO with steam is further converted to CO_2 and hydrogen through the Water Gas Shift Reaction. The details are explained below.

4.1.1 Coal Gasification

Coal gasification dates back to the mid-19th century, when it was used to make "town gas" for local cooking, heating, and lighting—many of the uses that natural gas meets today. Gasification works by mixing coal with limited oxygen, air, or steam at very high temperatures without allowing combustion to occur, called partial oxidation.

A gasification process is different from combustion. In gasification, carbon is partially oxidised to give carbon monoxide and hydrogen, opposite to carbon dioxide and water in combustion. Gasification occurs in the presence of limited oxygen, whereas combustion occurs in a stoichiometric excess of oxygen. The conversion of different elements of coal in the gasification and combustion processes is given in Table 4.

▼ Table 4: Conversion of coal elements on gasification and combustion

Gasification	Versus			Combistion
CO	←	C	→	CO_2
H_2	←	H	→	H_2O
N_2	←	N	→	NO_x
H_2S	←	S	→	SO_x
	←	O	→	O_2

The chemical reactions in gasification can progress to different extents depending on the gasification conditions (such as temperature and pressure) and the feedstock used. In comparison with conventional combustion, which uses a stoichiometric excess of oxidant, gasification typically uses one-fifth to one-third of the theoretical oxidant. This partially oxidises the carbon feedstock. As a "partial oxidation" process, the major combustible products of gasification are carbon monoxide and hydrogen, with only a minor portion of the carbon completely oxidised to carbon dioxide (CO_2). The heat produced by the partial oxidation provides most of the energy required to drive the endothermic gasification reactions.

Within a gasification process, the major chemical reactions involving carbon (C), carbon monoxide (CO), carbon dioxide (CO_2), hydrogen (H_2), steam (H_2O), and methane (CH_4) are as follows:
The combustion reactions:

1. $C + ½ O_2 \rightarrow CO$ (-111 MJ/kmole)
2. $CO + ½ O_2 \rightarrow CO_2$ (- 283MJ/kmole)
3. $H_2 + ½ O_2 \rightarrow H_2O$ (- 242MJ/kmole)

Other important gasification reactions include:

4. $C + H_2O \leftrightarrow CO + H_2$ (+131 MJ/kmole) Water Gas Reaction
5. $C + CO_2 \leftrightarrow 2CO$ (+172 MJ/kmole) Boudourd Reaction
6. $C + 2H_2 \leftrightarrow CH_4$ (-75 MJ/kmole) Methanation Reaction

The above combustion reactions (reactions 1 to 3) are essentially carried out to completion under normal gasification operating conditions. Under conditions of high carbon conversion, the 3 heterogeneous reactions (reactions 4 to 6) can be reduced to 2 homogeneous gas-phase reactions—namely, the water gas shift and steam methane reforming reactions (reactions 7 and 8) below. Together, these reactions play a key role in determining the final equilibrium composition of synthesis gas (syngas).

7. $CO + H_2O \leftrightarrow CO_2 + H_2$ (-41MJ/kmole) Water Gas Shift Reaction.
8. $CH_4 + H_2O \leftrightarrow CO_2 + 3H_2$ (+206 MJ/kmole) Steam Methane Reforming Reaction.

A positive sign in the above reactions means that it is endothermic and heat energy is supplied to carry out the reaction. A negative sign means the reaction is exothermic and heat is released during the reaction.

There are mainly five-unit steps in producing hydrogen from coal gasification, which are:

- Step 1 - Gasification: Gasification turns coal into a very hot (typically above 700°C) synthesis gas, or syngas, which is composed of carbon monoxide, hydrogen, and carbon dioxide, as well as small amounts of other gases and particles as impurities. This is accomplished by mixing pulverised coal with an oxidant, usually steam, oxygen, or both.
- Step 2 – Cooling and Cleaning: The syngas is cooled and cleaned to remove the other gases and particles, leaving only carbon monoxide, carbon dioxide, and hydrogen. Syngas from the gasification process is easier to clean due to the higher partial pressure than the flue gas emissions from a pulverised coal power plant. During syngas cleaning, mercury, sulphur, trace contaminants, and particulate matter are removed.
- Step 3 - Shifting: The syngas is then sent to a shift reactor. During the shift reaction, the carbon monoxide mixed with steam is converted into hydrogen and carbon dioxide. The product gas from the shift reactor consists of hydrogen and carbon dioxide.
- Step 4 - Purifying: Once the syngas has been shifted to CO_2 and H_2, it is purified by separating into the streams of hydrogen and carbon dioxide. The hydrogen stream is further refined to the required level of purity for use. The carbon dioxide stream is captured and sent off for sequestration.
- Step 5 - Using: There is now a stream of pure hydrogen ready for a variety of uses. It can be burned in a gas turbine to generate electricity, converted into electricity in a fuel cell, used as a fuel for an internal combustion engine, or used as a chemical to make fertiliser, semiconductors, and many other products. For use in a fuel cell, H_2 purity should be to the required level to avoid catalyst poisoning.

A flow sheet of coal gasification is given in Figure 10.

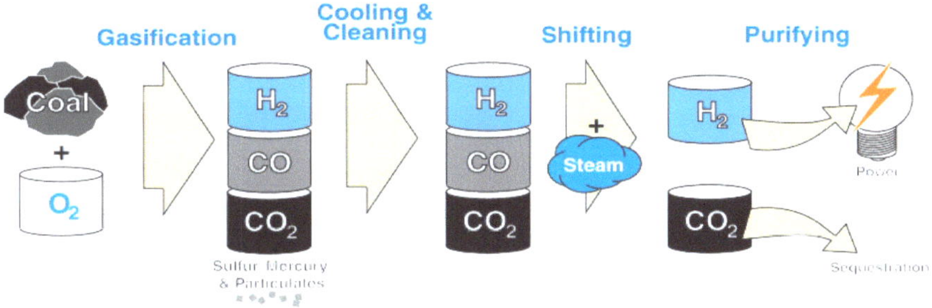

▲ Figure 10: Flow sheet of Coal Gasification

Currently, in India, only 2 commercially operated coal-based power plants use efficient and cleaner gasification technology using syngas to drive the turbine. These are located at Vindhyachal in Madhya Pradesh and at Singrauli in Uttar Pradesh; both are operated by NTPC. Gasification plants, in general, have better emissions profiles than conventional pulverised coal power plants for different types of emissions. Higher operating efficiencies in the gasification plants allow for significant reductions in pollutants.

4.2 Hydrogen from Oil and Natural Gas

In today's energy supply system, electricity, gasoline, diesel fuel, and natural gas serve as energy carriers. These carriers are made by the conversion of primary energy sources, such as coal, petroleum, underground methane, and nuclear energy, into an energy form that is easily transported and delivered in a usable form by industrial, commercial, residential, and transportation end users. The sustainable energy supply system of the future features electricity and hydrogen as the dominant energy carriers. Hydrogen would be produced from a diverse base of primary energy feedstocks, using the resources and processes that are most economical or consciously preferred. Methods to produce hydrogen from natural gas are well developed and account for over 95% of all hydrogen produced in the US and 48% globally.

In its 2018 report, the IEA indicated that 78% of hydrogen is produced from oil and gas combined, 30% from oil, and 48% from gas. Hydrogen from oil is mostly produced in refineries during catalytic cracking, catalytic reforming, and dehydrogenation, and accounts for 22% of global hydrogen production from fossil fuels.

Natural gas (specifically methane) is often preferred over oil for hydrogen production, particularly through steam methane reforming, due to its higher hydrogen content, lower carbon content, and easier processing, which leads to higher hydrogen yields at lower costs.

4.2.1 Steam Methane Reformation (SMR)

Hydrogen (H_2) production from natural gas primarily involves a process called steam methane reforming (SMR), in which high-temperature steam reacts with methane to produce hydrogen and carbon monoxide. Carbon monoxide is further processed to produce additional hydrogen through a process called the Water Gas Shift Reaction.

Unit operations that are involved in the SMR process are as follows:

Reformation Process

- The feed stream, natural gas or methane, is purified to remove any contaminants like sulphur compounds and chlorides, which can poison the catalyst in use. This shall increase the life of the catalysts.
- Natural gas (primarily methane) is mixed with high-temperature steam (700°C–1,000°C) in the presence of a catalyst. The reaction is endothermic, meaning it requires a heat source to proceed. The pressure at which the reaction takes place is typically moderate, ranging from 3–25 bar (43.5–362.5 psi).
- The reaction produces hydrogen (H_2), carbon monoxide (CO), and a small amount of carbon dioxide (CO_2).
- The chemical equation for the SMR reaction is:

$$CH_4 + H_2O \rightarrow CO + 3H_2.$$

Water Gas Shift Reaction:

- The carbon monoxide (CO) produced in the SMR process is further reacted with steam in a water gas shift (WGS) reactor. The WGS reaction, typically conducted as a two-stage process, involves high-temperature shift (HTS) at 310–450°C and low-temperature shift (LTS) at 200–250°C, with pressures ranging from atmospheric to 84 bar. The WGS reaction is exothermic. This means the reaction releases heat as it proceeds, favouring lower temperatures for higher conversion rates.
- This reaction converts CO into CO_2 and produces additional hydrogen. The chemical equation for the WGS reaction is $CO + H_2O \rightarrow CO_2 + H_2$.

Purification:

- The resulting mixture of hydrogen and CO_2 is then purified to obtain high-purity hydrogen. CO_2 and impurities are removed from the hydrogen stream using methods such as pressure-swing adsorption (PSA), liquid absorption, and the methanation process. Newer SMR plants utilise a PSA unit to produce 99.99% pure product hydrogen.
- In a liquid absorption system for CO_2 traces removal, the product gas from the shift reactor undergoes a methanation step to remove residual traces of carbon oxides.

High to ultra-high-purity hydrogen may be needed for the durable and efficient operation of fuel cells. Impurities cause various problems in the current state-of-the-art fuel cell designs, including catalyst poisoning and membrane failure. As such, additional process steps may be required to purify the hydrogen to meet industry quality standards.

Steam reforming of natural gas offers an efficient, economical, and widely used process for hydrogen production. The efficiency of the steam reforming process is about 65% to 75%, among the highest of current commercially available production methods. Natural gas is a convenient, easy-to-handle feedstock with a high hydrogen-to-carbon ratio. Methane is widely available from various sources. The cost of hydrogen produced by SMR is highly dependent on natural gas prices and is currently the lowest among all bulk hydrogen production technologies. A well-developed natural gas infrastructure exists all over the world, a key factor that makes hydrogen production from natural gas attractive.

Challenges of the SMR Process

During the production of hydrogen, CO_2 is also produced. To avoid the emission of CO_2, it can be captured and sequestered.

Sequestration technologies are relatively new, and there is no long-term evidence to demonstrate that they will be successful. Sequestration in oceans is controversial because of the possible adverse impact on the aquatic environment due to the reduced pH of water.

Steam reforming of natural gas is widely used in industry today. Hydrogen is produced by the SMR process in large centralised industrial plants for use in numerous applications, including chemical manufacturing and petroleum refining. R&D programmes are currently investigating the development of small-scale SMR technologies to enable the development of small-scale, distributed hydrogen production and delivery infrastructure.

The process flow diagram of the SMR process is shown in Figure 11.

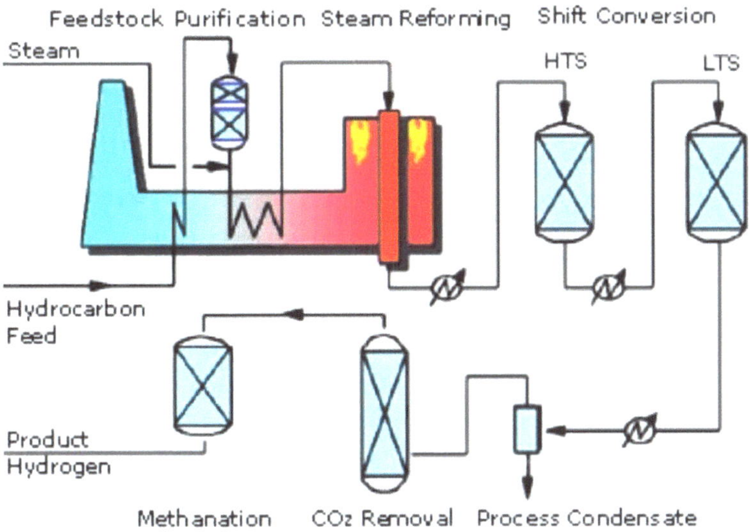

▲ Figure 11: SMR Process for hydrogen production

4.3 Production of Hydrogen from Biomass.

The conversion of biomass can take place *via* two routes: the thermogenic pathway (pyrolysis, gasification) or the biogenic pathway (anaerobic fermentation, alcoholic fermentation). Both approaches can produce gaseous and liquid fuels, with the first having high energy requirements. Biochemical transformation, on the other hand, proceeds essentially in a self-determined manner and with a lower external energy requirement.

These 2 pathways of producing hydrogen from biomass are discussed below separately.

4.3.1 Thermogenic Pathway

Thermogenic pathways for hydrogen production from biomass involve using heat and chemical reactions to convert biomass into valuable products, including hydrogen.

Biomass Gasification

Biomass gasification is similar to coal gasification, in which a synthetic gas (a mixture of hydrogen and carbon monoxide) is produced by partial oxidation of biomass with oxygen and steam. Carbon monoxide is further converted to hydrogen through the WGS reaction.

A simplified example reaction of biomass gasification is:

$$C_6H_{12}O_6 + O_2 + H_2O \rightarrow CO + CO_2 + H_2 + \text{other species}$$

The above reaction uses glucose as a surrogate for cellulose.

Actual biomass has a highly variable composition and complexity, with cellulose being one major component, while others are hemicellulose and lignin. Cellulose and starch are not sugars themselves, but rather Error! Hyperlink reference not valid. made of long, chain-like polymers of sugar units. While they are built from glucose, they are not considered «simple sugars» because they are not sweet, do not dissolve easily, and are not immediately absorbed by the body like table sugar.

Hydrogen is produced by reacting carbon monoxide with water; this process is called the water gas shift reaction.

$$CO + H_2O \rightarrow CO_2 + H_2 \text{ (+ small amount of heat released)}$$

Biomass gasification is a mature technology that uses a controlled process involving heat, steam, and oxygen to convert biomass to hydrogen and other products. Because growing biomass removes carbon dioxide from the atmosphere, the net carbon emissions of this method can be negative when coupled with CCUS.

Biomass, a renewable organic resource, includes agricultural crop residues (such as corn stover or wheat straw), forest residues, special crops grown specifically for energy use (such as switchgrass in North America, willow trees in the Ladakh region, and Jatropha), organic municipal solid waste, and animal wastes.

Plants consume carbon dioxide from the atmosphere during natural growth to produce biomass, offsetting the carbon dioxide released from producing hydrogen through biomass gasification and resulting in no net greenhouse gas emissions.

Here's a more detailed breakdown of hydrogen production from biomass.

- Raw biomass (e.g., wood, agricultural residues, algae) is collected and stored.
- The biomass is required to be prepared, such as chopping, grinding, or drying, to improve its suitability for the next stage.
- Biomass is heated in the absence of oxygen to produce bio-oil, char, and gases, including hydrogen. Biomass is heated to even higher temperatures in a controlled environment (using steam or oxygen) to produce a mixture of gases, including hydrogen, carbon monoxide, and carbon dioxide, known as syngas.

- The syngas is cooled and condensed to remove tars and other impurities.
- Carbon monoxide is converted to carbon dioxide using a WGS reactor to produce additional hydrogen and carbon dioxide.
- Carbon dioxide is removed from the syngas to produce a high-purity hydrogen stream.
- The purified hydrogen gas is compressed and stored for later use.
- The compressed hydrogen can be transported via pipelines or other means for various applications.

In the pyrolysis process (heating in the absence of hydrogen), gas, char, and bio-oil are formed. Char and bio-oil can be diverted to the gasification route to produce hydrogen, or char can be used as fuel or as an adsorbent, and bio-oil can be used as a feedstock for many value-added products.

It is worth mentioning here that Gensol Engineering Limited, in partnership with Matrix Gas & Renewables Ltd, has secured a contract worth ₹164 crores to develop India's first biohydrogen project, as reported by *Energyworld*. The project aligns with India's NGHM and marks a significant step towards sustainable energy. This will convert 25 tonnes of bio-waste into 1 tonne of hydrogen per day. Gensol and Matrix have partnered with Westinghouse, USA, to utilise their patented Pre-Gasification Plasma Induced Radiant Energy-Based Gasification System (GH2-PREGS) technology for the project.

Besides the Gensol project, there are other initiatives in India focused on biohydrogen production, such as a plant in Assam and another in Uttar Pradesh. These projects highlight the growing interest and potential of biohydrogen in India.

This biohydrogen plant is a key step in India's push towards sustainable energy and a circular economy. It aligns with the NGHM, which aims to promote green hydrogen production and usage.

A simplified block diagram, taken from IEA Bioenergy, of obtaining hydrogen from biomass gasification is given in Figure 12.

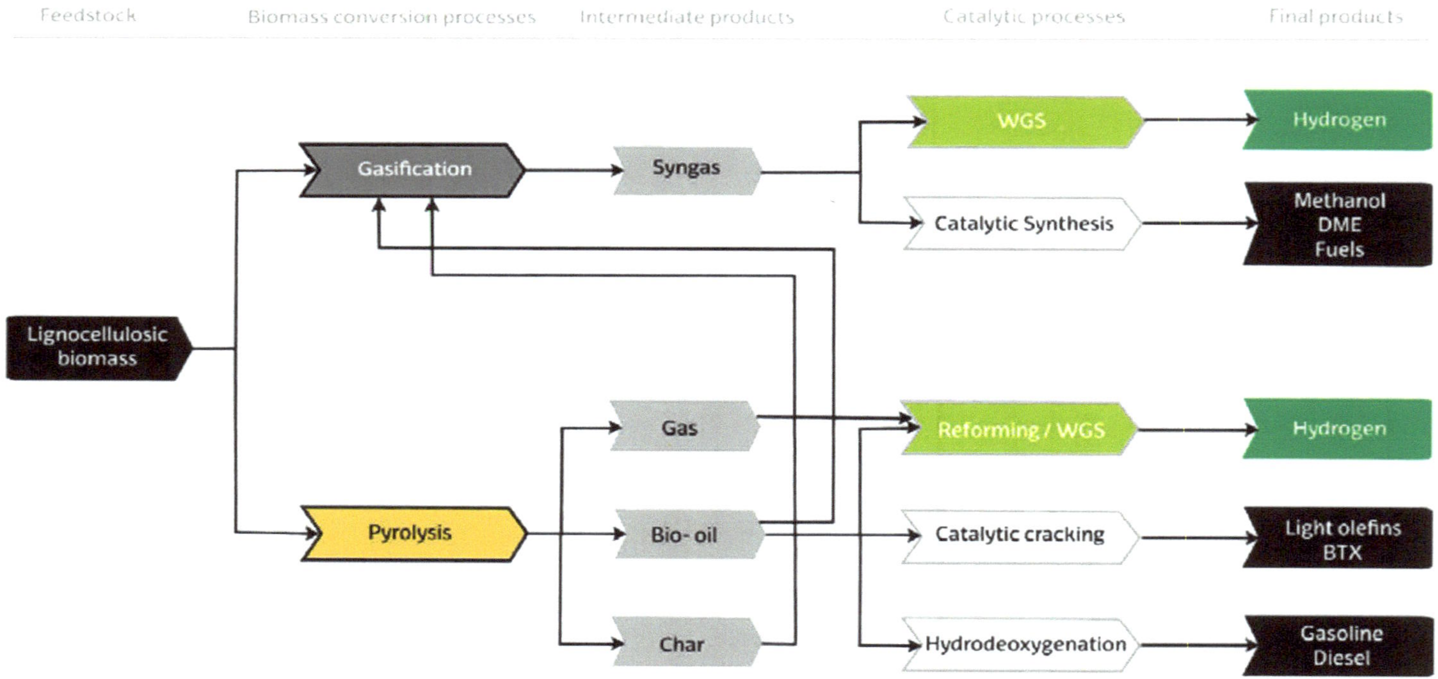

▲ Figure12: Schematic representation of the main processes involved in a thermochemical conversion route, based on lignocellulosic biomass (Arregi et al., 2018)

4.3.2 Biogenic Pathway.

The biogenic pathway does not involve external heat for conversion. Specific bacteria convert carbohydrates to hydrogen in steps. Hydrogen can be produced through biogenic processes, which involve using biological systems like microorganisms to convert organic matter into hydrogen. These processes are generally classified as either biophotolysis or dark fermentation, with the former utilising sunlight and microorganisms like algae and cyanobacteria to split water, and the latter employing bacteria to break down organic matter in the absence of sunlight. Both are carried out under anaerobic conditions.

Biophotolysis is a sustainable biological process that uses solar energy and microorganisms (microalgae or cyanobacteria) to split water into hydrogen and oxygen as a clean fuel production method, primarily categorized into direct biophotolysis (using water directly) and indirect biophotolysis (using stored carbohydrates).

- Mechanism: Photosynthetic organisms (e.g., *Chlamydomonas reinhardtii*) use Photosystem to capture light energy, breaking water molecules to produce electrons, which are then used by enzymes like hydrogenase or nitrogenase to create hydrogen.
- Direct Biophotolysis**:** Green algae directly split water into hydrogen and oxygen. However, the oxygen produced often inhibits the hydrogenase enzyme.
- Indirect Biophotolysis**:** Cyanobacteria produce hydrogen through a two-stage process: first by accumulating biomass (carbohydrates) via photosynthesis, and second by breaking down that biomass to produce hydrogen, which allows for separating oxygen production from hydrogen production, mitigating the inhibition problem.

Dark fermentation is a biological process performed in the absence of light and oxygen, where microorganisms break down organic matter to produce hydrogen, often utilising waste materials as a substrate. Dark fermentation can utilise a wide range of organic substrates, including carbohydrates, proteins, and even waste materials. Hydrogen production rates can be higher compared to photofermentation. Besides hydrogen, other products like carbon dioxide, organic acids, and alcohols are produced.

The biochemical reaction taking place in the dark fermentation is given below:

$$C_6H_{12}O_6 + 2H_2O \rightarrow 2CH_3COOH + 2CO_2 + 4H_2$$

Fermentation and biophotolysis are two popular methods used in biohydrogen production (refer Figure 13).

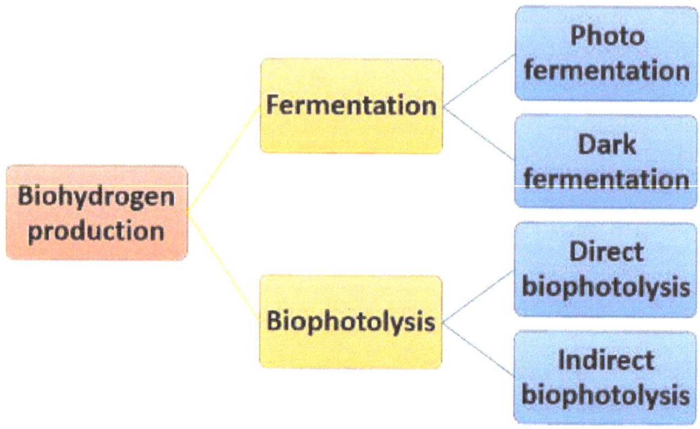

▲ Figure 13: Hydrogen production from biogenic pathway

Advantages of biological hydrogen production

- It utilises renewable resources like sunlight and biomass.
- It can be used to treat organic waste, providing a dual benefit of energy production and waste management.
- If sourced from certain waste materials, it can be carbon neutral or even carbon negative, meaning it can remove more carbon from the atmosphere than it releases.
- Biological processes can potentially reduce the cost of hydrogen production, especially when utilising waste materials, according to the European Biogas Association.

Key considerations:

- Yield and Rate: While biohydrogen production is promising, the yield and rate of hydrogen production can be low compared to other methods.
- Feedstock Costs: The cost of suitable biomass or organic waste can be an important factor.
- Process Optimisation: Research is ongoing to optimise biological processes to improve hydrogen yield, rate, and cost-effectiveness.
- Research focuses on developing oxygen-tolerant enzymes, optimizing light-dark cycles, and creating better bioreactor designs to enhance production efficiency.

Biomass-to-hydrogen (biohydrogen) is one of the most scalable routes to low emission, and potentially carbon-negative. Important factors that potential investors and project developers care about, are given below. These figures are indicative, reported in literature.

- Yield: ~100 kg H_2 per tonne of dry wood/straw biomass
- Energy efficiency: typically, ~40–70% (LHV basis). Biomass energy efficiency varies by conversion method, with traditional electricity generation via steam, converting roughly 30% of feedstock energy, while advanced gasification can reach up to 60%, and further with heat recovery it can go to 70%.
- Technology maturity: TRL (Technical Readiness Level) ~5–7 (integrated full-chain scale-up still needed).
- Cost signal: a large-scale (200 MW) plant is estimated around ~€4/kg H_2 at a biomass price of €20/MWh, with potential reductions to <€3/kg with process improvements and CCS.
- Carbon intensity (LCA): median ~4.7 kg CO_2e/kg H_2. With CCS studies an average around −15.8 kg CO_2e/kg H_2 (net-negative range) has been reported.

Currently, biological systems reach about 20%–30% of the theoretical energy yield (the "Thauer limit" of 4 moles H2 per mole of glucose). While energy efficiency in biogenic processes is lower than electrolysis, the resource efficiency is superior for organic waste. It converts "negative value" feedstocks (slurries, agro-waste) into hydrogen.

In 2030 Outlook, a two-stage biogenic process, dark fermentation followed by methanogenesis, is becoming very attractive, maximizing energy extraction from biomass by producing both H2 and CH4.

Field Comparison of Electrolysis and Dark Fermentation, sourced from IEA, IRENA, Hydrogen Council- hydrogen Insights, is given below.

▼ Table 5: Comparison Between Electrolysis and Dark Fermentation

Metric	Electrolysis (PEM)	Dark Fermentation
System Efficiency	60% – 70%	20% – 30% (substrate energy)
Input Requirement	Renewable Electricity	Organic Waste / By-products
TRL (Tech. Readiness)	8-9 (Commercial)	4-6 (Pilot/Demo)
Scalability	High (Modular)	Medium (Feedstock limited)
Main By-product	High-purity Oxygen	Organic acids / Biogas

Electrolysis wins on a purity scale which is essential for Fuel Cell and heavy industries. However, for the agricultural and waste-treatment sectors, Dark Fermentation represents a more sustainable "circular" efficiency. The real winner isn't a single technology, but the hybridization of these systems within energy hubs.

4.4 Methane Pyrolysis

Methane pyrolysis is a low-carbon, endothermic process that thermally decomposes natural gas to produce «turquoise» hydrogen and solid carbon, avoiding direct emissions. It requires high temperatures (800 to 1600°C) and offers a potentially cost-competitive, lower-emission alternative to steam methane reforming, with solid carbon as a saleable byproduct.

The reaction breaks C-H bonds in methane without oxygen, producing hydrogen gas and solid carbon (graphite or carbon black)

Techniques to achieve the necessary high temperatures include thermal, plasma, molten metal/salt, and catalytic cracking.

It is highly efficient, has low water usage, and generates solid carbon, which is easier to handle than CO_2 storage.

The process is energy-intensive, requires high temperatures, faces catalyst deactivation (coking), and needs a stable market for the produced solid carbon to be economically viable.

Companies working on commercialising this technology include Monolith, HazerGroup, and Modern Hydrogen.

C-Zero›s process, a version of methane pyrolysis, was developed and demonstrated by Eric McFarland, a chemical engineering professor at the University of California, Santa Barbara. Backed by funding from the US Department of Energy, along with investors like Bill Gates' Breakthrough Energy Ventures, C-Zero is working to commercialise its process for large-scale clean hydrogen production. A pilot facility, capable of producing 400 kg of hydrogen per day, was supposed to be operational. US policy on climate change will decide the future of such an innovative approach, unless private players like Bill Gates come forth.

5 Water Electrolysis

Water electrolysis is the second most widely practised pathway for producing hydrogen, after SMR and coal gasification. Though at present, less than 2% of global hydrogen comes from electrolysis, it is growing at a much faster rate due to the global decarbonisation campaign, cheaper renewable power, and more efficient electrolysers. Electrolysis has been used for many years to generate hydrogen for the ammonia synthesis process.

The technology to produce hydrogen from water electrolysis is discussed in detail below.

5.1 Working Principle of Electrolysis.

Electrochemistry and thermodynamics work together in electrolysis. Water electrolysis is an electrochemical reaction where ions and electrons flow by splitting water into hydrogen and oxygen. Electrochemical reaction in the electrolytic cell is just the opposite of the galvanic cell. In an electrolytic cell, electrons are made to flow through the application of DC electric current, whereas in a galvanic cell, electrons flow spontaneously once the circuit is completed. A galvanic cell converts chemical energy to electric energy, whereas an electrolytic cell converts electric energy to chemical energy in the form of hydrogen gas.

An electrochemical cell consists of 4 basic elements: anode, cathode, conductor, and electrolyte. Oxidation takes place at the anode, where electrons are released, and reduction takes place at the cathode, where electrons are consumed. Both oxidation and reduction reactions, called redox reactions, take place simultaneously. The oxidation half-reaction occurs at the anode, and the reduction half-reaction occurs at the cathode. Both half reactions make a full redox reaction. The electrodes are connected by a conductor and an electrolyte. An electrolyte is an ionic substance or solution that contains equal negative and positive ions, thereby maintaining the system's electrical neutrality.

Electrons flow through conductors and ions through electrolytes. Negative ions, called anions, move towards the anode. Positive ions, called cations, move towards the cathode. Electrolysis is a non-spontaneous reaction, meaning that the reaction will proceed only when energy is supplied, whereas a galvanic cell reaction is spontaneous as it needs no energy to initiate the reaction.

Figure 14 shows both types of electrochemical cells—galvanic cell and electrolytic cell—with half reactions occurring at the anode and cathode.

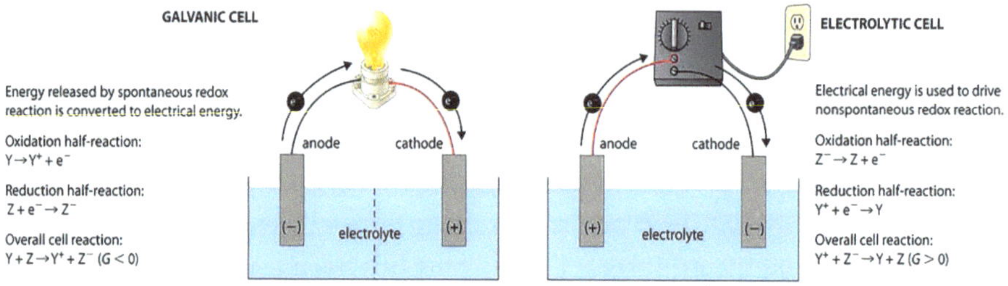

▲ Figure 14: Electrochemical Cell

In the electrolytic cell above (right side), an external source of electrical energy is used to generate a potential difference between the electrodes that forces electrons to flow, driving a non-spontaneous redox reaction; only a single compartment is employed in most applications. Electrolytic cells are called electrolysers. In both kinds of electrochemical cells, the anode is the electrode at which the oxidation half-reaction occurs, and the cathode is the electrode at which the reduction half-reaction occurs.

5.2 Thermodynamics of the Water Electrolysis

Thermodynamic Functions of State

Liquid water can be dissociated into its elemental components, molecular hydrogen and oxygen, according to the equation:

$$H_2O(l) \rightarrow H_2(g) + 1/2\ O_2(g) \tag{1}$$

In standard conditions of temperature and pressure ($T°$=298 K, $P°$=1 bar), water is liquid and H_2 and O_2 are in the gaseous state. Standard change of enthalpy of dissociation($\Delta Hd°$), Standard change of entropy of dissociation ($\Delta Sd°$), and Standard change of free Gibbs energy of dissociation ($\Delta Gd°$) are related by the following equation, as per the second law of thermodynamics:

$$\Delta Hd° = \Delta Gd° + \mathrm{T}\Delta Sd° \tag{2}$$

For the above reaction (1), these thermodynamic values at the standard condition of temperature $T°$ (298 K) and pressure P (1 bar) are

$\Delta Hd°(H_2O(l)) = +285.840$ kJ per mol.
$\Delta Sd°(H_2O(l)) = +163.15$ J per mol per K.

At 298 K, the change of entropy at standard pressure (1 bar) is 163.15 multiplied by 298, which is equal to 48618.7 J per mole or 48.618 kJ per mole.

$$\Delta Gd°(H_2O(l)) = \Delta Hd°(H_2O(l)) - T \cdot \Delta Sd°(H_2O(l)) = +237.22 \text{ kJ per mol}$$

In spite of a favourable entropic contribution due to the formation of 1.5 moles of gaseous species, as per equation (1), the enthalpy change is positive; as a result, it is an endothermic reaction. The Gibbs Free Energy change is positive, and the reaction is non-spontaneous.

Water vapour can also be dissociated into gaseous hydrogen and oxygen according to:

$$H_2O(vap) \rightarrow H_2(g) + 1/2\ O_2(g) \qquad (3)$$

Enthalpy, entropy, and free (Gibbs) energy standard changes for reaction (3) are, respectively:

$\Delta Hd°(H_2O(vap)) = +241.80$ kJ per mol
$\Delta Sd°(H_2O(vap)) = +44.10$ J per mol per K
$\Delta Gd°(H_2O(vap)) = \Delta Hd°(H_2O(vap)) - T \cdot \Delta Sd°(H_2O(vap)) = +228.66$ kJ per mol

From the values of the state functions of water, liquid, and water vapor dissociation equations (1) and (3), respectively.

$\Delta Hd°(H_2O(l)) - \Delta Hd°(H_2O(vap)) = +44.04$ kJ per mol is the difference of enthalpy change due to water vaporisation at standard conditions.

$$H_2O(l) \rightarrow H_2O(vap) \qquad (4)$$

Entropy change due to water vapourisation in standard conditions of equation (4) above is

$\Delta Sd°(H_2O(l)) - \Delta Sd°(H_2O(vap)) = +119.05$ J per mol per K is the difference of entropy change due to water vapourisation.

The state functions, change of enthalpy, Gibbs Free Energy, and entropy at different splitting temperatures and one bar pressure are given below in Figure 15. A plot indicating the total energy required, both electrical energy and heat energy, to produce hydrogen by splitting the water through the electrolytic process given below.

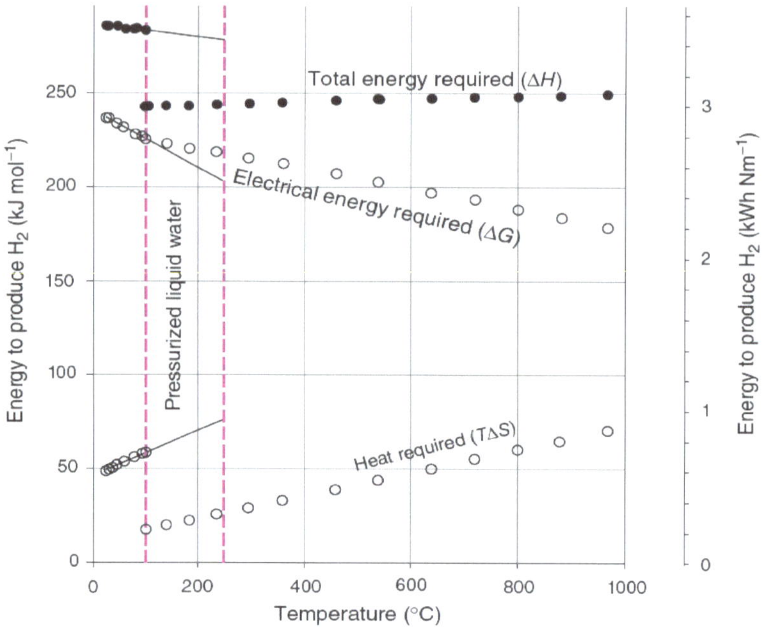

▲ Figure 15: $\Delta G(T)$, $\Delta H(T)$, and $T\Delta S(T)$ of the water splitting reaction at P = 1 bar. (—) data for pressurized liquid water up to 250°C.

It can be seen in Figure 15 that state functions like change of state functions at temperature T, say $\Delta G(T)$, $\Delta H(T)$, and $T\Delta S(T)$ for reaction (1) change with temperature. The total energy ΔH (T) required to split 1 mole of water is almost constant over the entire temperature range of practical interest. The entropy change ΔS is also approximately constant and positive, but the entropic contribution ($T \cdot \Delta S$) increases with temperature. The Gibbs Free Energy change $\Delta G = \Delta H - T \cdot \Delta S$ is positive but decreases as the operating temperature increases. ΔG becomes negative at high temperatures (>2,500 K). When ΔG becomes negative, the splitting of water needs no external energy. It splits by itself at such a high temperature.

Few, if any, materials can sustain such high temperatures more than 2,500 K and, therefore, the direct thermal dissociation of water is not considered for practical applications. Discontinuities on ΔH (100°C, 1bar) and $T{\cdot}\Delta S$ (100°C, 1bar) observed at 100°C are due to water vapourisation. The magnitude of the enthalpy discontinuity or decrease is equal to the standard enthalpy of water vapourisation (≈ +45kJ per mol) and equal to the decrease of the entropic contribution. The slope difference of $T \cdot \Delta S(T)$, before and after 100°C, reflects the entropy change after water vapourisation. In Figure

15, solid lines between 100 and 250°C show the situation for the electrolysis of liquid water above 100°C (pressurised liquid water).

In conclusion, a higher operating temperature facilitates the dissociation of water by decreasing the electrolysis voltage. At room temperature, 15% of the total energy required for electrolysing water comes from heat and 85% from electricity. At 1,000°C, one-third comes from heat and two-thirds from electricity.

This is why high-temperature water electrolysis is interesting when heat is available. High-temperature electrolysis requires less electricity (which is more expensive than heat).

Selection Criteria for Operating Temperature

Water electrolysis can be performed at different operating temperatures. The question is, what is the best temperature?

ΔH (the total amount of energy required to split 1 mole of water) is almost temperature independent (see Figure 15), but the higher the operating temperature, the less electricity is required, and the more heat can be used. The first criterion used to select the operating temperature is the cost ratio of kilowatt-hours of electricity to kilowatt-hours of heat, which is usually between 3 and 5.

5.3 Electrochemistry of Electrolysis

Thermodynamics tells us that the water-splitting reaction is non-spontaneous because its Gibbs Free Energy change is positive. Therefore, the reaction can be driven by externally providing energy to the system in the form of electricity. This is a so-called endergonic transformation, and the device used in the process is an electrolyser. An electrolyser contains at least one electrolysis cell. An electrolytic cell has 2 electrodes placed face-to-face and separated by a thin layer of an ionic conductor (the electrolyte). Electrical work is provided to the cell to split water molecules into gaseous hydrogen and oxygen. Half-cell reactions (and associated mechanisms) depend on electrolyte pH. In acidic media, water splitting occurs according to:

$$\text{Anode}: H_2O(l) \rightarrow 1/2\ O_2(g) + 2H^+ + 2e^- \quad (5)$$

$$\text{Cathode}: 2H^+ + 2e^- \rightarrow H_2(g) \quad (6)$$

$$\text{Full reaction is}: H_2O(l) \rightarrow H_2(g) + 1/2\ O_2(g) \quad (7)$$

Equation (4) is the oxygen evolution reaction (OER), and equation (5) is the hydrogen evolution reaction (HER). In terms of mass balance, there is a net consumption of water molecules at the anode and production of gases at each of the two-electrode/electrolyte interfaces. From an energy viewpoint, according to the first principle of thermodynamics, the amount of electricity equivalent to nFE is required at equilibrium to split 1 mole of liquid water, according to equation (7), which is equal to the Gibbs Free Energy change (ΔGd) of water dissociation reaction:

$$\Delta Gd - \mathrm{nFE} = 0, \text{ where } \Delta Gd > 0 \text{ in water dissociation} \tag{8}$$

where $n = 2$ is the number of electrons exchanged during the electrochemical splitting of one water molecule; $F \approx 96{,}485$ Coulombs per mol is the Faraday Constant, that is, the electric charge of 1 mole of electrons; and E is the electrolysis potential associated with reaction (7) in terms of volts. ΔGd (in J per mol) is the free energy change associated with reaction (7).

Water dissociation occurring at constant temperature T and pressure P, one can write:

$$\Delta Gd\,(T, P) = \Delta Hd\,(T, P) - T\,\Delta Sd\,(T, P) \tag{9}$$

$\Delta Hd\,(T, P)$, $\Delta Sd\,(T, P)$, and $\Delta Gd\,(T, P)$ are, respectively, the enthalpy change, entropy change, and the Gibbs Free Energy change per mol of reaction (7) occurring at temperature T and pressure P. As shown in Figure 15, ΔGd (in the form of electricity) and $\mathrm{T}\cdot\Delta Sd$ (heat contribution) are required to split 1 mole of water, in KJ/Mole of water.

Two different thermodynamic voltages are usually used to characterise water electrolysis.

The free energy electrolysis voltage E in volts, called Theoretical or Reversible or Equilibrium Voltage at standard temperature and pressure, is defined as:

$$E\,(T, P) = \Delta Gd\,(T, P) \,/\, n\,F \tag{10}$$

The enthalpy or thermoneutral voltage V in volts is defined as:

$$V\,(T, P) = \Delta H\,(T, P) \,/\, n\,F \tag{11}$$

Reference electrolysis voltages are usually calculated in standard conditions of temperature ($T°= 298$ K) and pressure ($P°=1$ bar). Water is liquid, H_2 and O_2 are gaseous, and standard free energy, enthalpy, and entropy changes for reaction (7) are:

$$\Delta Gd°(H_2O(l)) = +237.22 \text{ kJ per mol}$$
$$E° = \Delta Gd°(H_2O(l))/2F = 237.22 \,/\, 2\text{x}96.485 = 1.23 \text{ V}$$

(one Joule represents energy transferred when one Coulomb moves through a potential difference of one Volt.)
Eo above in voltage is called the theoretical voltage.

$\Delta Hd°(H_2O(l)) = +285.840$ kJ per mol.
$V° = \Delta Hd°(H_2O(l))/2F = 285.84 / 2x96.485 = 1.48$ V.

The voltage, V_0, above is called the thermoneutral voltage.

E_0 (in volts) is the free energy electrolysis voltage associated with reaction (7) at standard conditions of temperature and pressure.

ΔGd_0 (in kJ per mol) is the free energy change associated with reaction (7) at standard temperature and pressure.

V_0 is the enthalpy or thermoneutral electrolysis voltage at standard conditions of temperature and pressure associated with reaction (7). Thermoneutral means no heat energy is supplied from an external source to split. The heat energy required to split is drawn from electrical energy.

E_0 is called the theoretical, reversible, or equilibrium voltage. V_0 is called the thermoneutral voltage.

5.4 Understanding Voltage in an Electrolytic Cell:

Electrolysis is the process of using electrical energy to drive non-spontaneous electrochemical reactions. To understand the voltage requirements for this process, it's essential to differentiate between theoretical, thermodynamic, thermoneutral, and overvoltage.

Theoretical Voltage (Standard Cell Potential, E°cell)

- This represents the minimum voltage required to initiate an electrolytic reaction under standard conditions (25°C, 1 atm pressure, 1 M concentration for solutions). Theoretical voltage of water dissociation is 1.23 volts. This is an open-circuit voltage and was worked out on the basis of the change in only the Gibbs Free Energy in water dissociation, equation (10) above. This is also called Reversible Voltage or Equilibrium Voltage.
- It is calculated by subtracting the standard electrode potential of the anode from that of the cathode: E°cell = E°cathode - E°anode.
- A negative E° cell indicates that the reaction is non-spontaneous (opposite to Gibbs Free Energy change, ΔG) and requires an external energy input (like a battery) to proceed, according to MCAT-Review.org.

Thermodynamic Voltage

The thermodynamic voltage of water electrolysis at standard conditions (25°C, 1 atm) is 1.23 V, which represents the minimum electrical energy needed to split water into hydrogen and oxygen at standard conditions of 298 Kelvin and 1 atm. This voltage is defined by the Gibbs Free Energy change of the reaction, by equation (10) above. The thermodynamic voltage at standard temperature and pressure is called the theoretical voltage. Thermodynamic voltage depends upon the temperature and pressure, and is related with Nerst Equation.

The required thermodynamic voltage decreases as temperature increases (1.18V at 100°C and 0.9V at 1000°C) as more energy input is provided by heat, rather than electricity.

Thermoneutral Voltage

The thermoneutral voltage of an electrolyser is the voltage at which the energy required for the non-spontaneous reaction is entirely supplied by electricity, without any net heat exchange with the surroundings (Adeabetic). This voltage is calculated on the basis of the total enthalpy change (ΔH) of the reaction, as per equation (11) above. It represents a higher energy input than the reversible voltage, as it includes the energy needed to heat the products to their formation temperature.

The thermodynamic voltage of water dissociation is 1.48 volts.

Irreversible Voltage (Overpotential Or Overvoltage)

- In practice, the applied voltage for electrolysis is always higher than the theoretical or thermodynamic voltage due to irreversible losses in the system.
- This additional voltage is called overpotential or overvoltage, and it represents the extra energy required to overcome various resistances and kinetic limitations.
- There are 3 main types of overpotential caused due to polarisation. In electrochemistry, polarisation is the deviation of an electrode's potential from its equilibrium value when an electric current flows, caused by slow electrode reactions or ion transport, creating a "back e.m.f." that opposes the current and hinders the process. It's categorised into activation polarisation (slow reaction kinetics), concentration polarisation (ion diffusion limits), and resistance polarisation (ohmic drop/insulating films).

- Activation overpotential: Related to the energy needed to initiate electrochemical reactions at the electrode surfaces.
- Concentration overpotential: Occurs due to changes in reactant and product concentrations at the electrode surface, limiting ion diffusion. This is calculated from the Nernst Equation.

It should be noted that the change in Gibbs Free Energy is a function of temperature and pressure, and changes in these parameters will change the free energy of the electrolysis voltage. For example, at 100°C, the reversible cell voltage drops to 1.18 V, while at 1000°C it drops to ≈ 0.9 V. Except in rare cases where there is a continuous supply of heat, this does not play a big role, so it can be assumed that low-temperature electrolysis will be carried out at a thermoneutral voltage above EThermoneutral.

The amount of energy required to drive the reaction in the cell does not take into account the rate at which the reaction occurs, which can be considered to be proportional to the current. In order to achieve the desired reaction rate, there must be an overpotential to overcome the resistance present in the system (such as the energy to overcome the activation energy barrier of the reaction) and to achieve these current densities.

The overpotential generated by the electrode reaction increases with increasing current density. The actual cell voltage is different from the reversible cell voltage due to system inefficiencies.

The actual cell voltage can be decomposed into:

$$\text{Ecell} = \text{Erev} + \eta\text{Anode} + \eta\text{cahode} + \text{IRcell} + \eta\text{MT}$$

where ηAnode is the overpotential at the anode, ηcathode is the overpotential at the cathode, I is the current, RCell is the resistance of the cell, and ηMT is the voltage inefficiency caused by mass transport limitations at the electrodes. Figure 16 shows an approximate decomposition of the voltage contribution for conventional alkaline electrolysis. It can be seen that the contribution of ohmic losses to the voltage begins to dominate at a current density of 500 mA·cm^2.

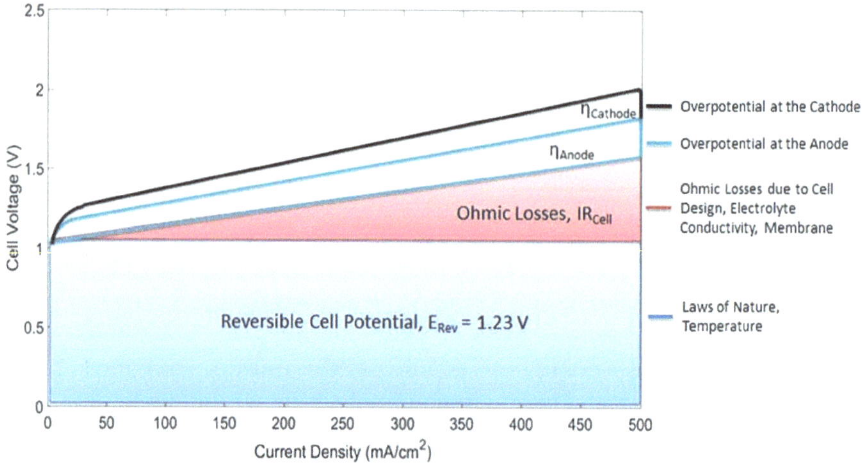

▲ Figure 16: Decomposition of Voltage Contribution in Alkaline Electrolysis

Rough breakdown of contributions to cell voltage. Mass transfer overpotential is not well defined.

In order to optimise the cell voltage, these influencing factors must be quantified and optimised separately to improve electrolysis efficiency.

5.5 Nernst Equation.

The Nernst Equation, given in Figure 17, tells us what the potential of an electrode is when the electrode is surrounded by a solution containing a redox-active species with an activity of its oxidised and reduced species.

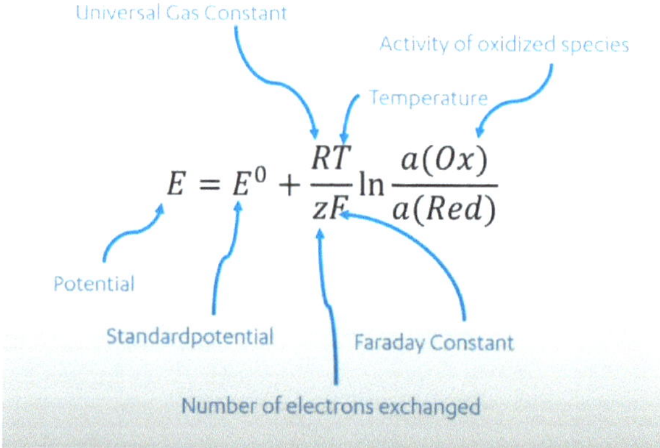

▲ Figure 17: Nerst Equation

E is the electrode potential of the half-reaction at the electrode (anode or cathode), and the activities of the reduced and oxidised species are a(Ox) and a(Red) respectively. The remaining parameters in the equation are the universal gas constant R, the absolute temperature T, the Faraday Constant F, the standard potential of the reaction Oxidation or Reduction (anode or cathode) E^0, and the number of transferred electrons per molecule z. The net potential or voltage of an electrolytic cell is the difference between the positive and negative electrodes, that is $E_{cell} = E_{cathode} - E_{anode}$.

A positive ΔG (non-spontaneous reaction) corresponds to a negative Ecell for electrolysis, according to the University of Texas at Austin.

Bubble Overpotential: Resistance and hindered surface area due to gas bubbles (H_2/O_2) forming on the electrode surface. It is not taken into account in Figure 17 above.

The magnitude of overpotential depends on factors like electrode material, surface characteristics (roughness, defects), current density, temperature, and electrolyte composition.

For example, typical operational cell voltages for alkaline electrolysers range from 1.85–2.05 V and for PEM electrolysers from 1.75–2.2 V. This means that the overpotential in these systems can be substantially higher than the theoretical voltage (1.23V) and thermodynamic voltage (1.48V).

In essence, while the theoretical minimum voltage to split water is 1.23V, practical water electrolysis systems require a higher applied voltage due to thermodynamic considerations and various inefficiencies (overpotential). Understanding and minimising these overpotential contributions are crucial for improving the efficiency and reducing the cost of hydrogen production through water electrolysis.

pH Dependence of Water Dissociation Voltage

The free energy electrolysis voltage (Eo) required to split water into hydrogen and oxygen is pH-independent. The only difference between alkaline and acidic media is that the potential of each electrode is shifted along the potential axis by equal magnitude, as a function of electrolyte pH, with no change in net potential difference between those two electrodes. The consequences of pH are mostly on electrode material stability.

In acidic media, most metals are corroded and, hence, PGMs. are needed. This usually involves unsupported platinum or carbon-supported platinum nanoparticles for HER and iridium dioxide for OER. In alkaline media, nickel and cobalt electrodes are passivated, and oxides/hydroxides are electrochemically active.

5.6 Types of Electrolysers

There are 4 main types of water electrolysers: Alkaline water electrolysis (AWE), proton exchange membrane (PEM) electrolysis, anion exchange membrane (AEM) electrolysis, and solid oxide electrolysis cell (SOEC). Each type utilises a different electrolyte and operates under varying conditions, affecting their efficiency and suitability for different applications.

Each type of electrolysers is discussed below in detail.

5.6.1 Alkaline Water Electrolysis (AWE):

Alkaline water electrolysis (AWE) is the most mature and cost-effective technology. It uses an alkaline electrolyte solution—often potassium or sodium hydroxide to enable the flow of ions between electrodes. This electrolyser, which typically operates at low temperatures (60–80°C), has been the standard for decades and is widely recognised for its robust design and affordability. It's known for its long-term reliability and lower initial investment, but typically has lower efficiency than other types.

Pros of Alkaline Electrolysers

- Lower Production Costs: Due to simpler materials and designs, production costs from alkaline electrolysers are lower than those of other types, making them more accessible.
- Durability: Alkaline electrolysers are known for their longevity and consistent performance over time.
- Proven Technology: Trusted in industrial hydrogen production, alkaline electrolysers are a well-known solution with a long operational history.
- Ease of Maintenance: With fewer complex components, maintenance is simpler and requires less specialised expertise.
- Scalability for Large-Scale Production: Alkaline electrolysers are well-suited to large setups where costs need to be controlled, allowing for steady hydrogen output.

Cons of Alkaline Electrolysers

- Lower Efficiency: Alkaline technology, in some cases, has lower energy efficiency, particularly at high current densities.
- Longer Startup and Shutdown Times: Alkaline electrolysers are slower to start and adjust their output, which can limit their effectiveness in applications with variable energy sources.

- Lower Hydrogen Purity: Compared to PEM electrolysers, alkaline electrolyser provides lower hydrogen purity due to possibilities of migration of ions. Additional H_2 purification steps may be required, especially for applications that require very high-purity hydrogen, adding to operational costs.

5.6.2 Proton Exchange Membrane (PEM) Electrolysis:

PEM electrolysers use a solid polymer membrane as an electrolyte, which allows only protons to pass through. This technology offers higher efficiency, higher current densities, and the ability to operate at higher pressures, making it suitable for larger-scale hydrogen production. However, it often relies on expensive platinum group metal catalysts.

PEM electrolysers represent a relatively newer approach to hydrogen production. At their core, they rely on a solid polymer electrolyte that separates the electrodes within the cell and facilitates only proton movement during the electrolysis process. The membrane, made of specially designed polymers, allows only positively charged ions to pass through, blocking gases like oxygen and hydrogen to keep the system stable and efficient.

Here's a breakdown of how PEM electrolysis works, step-by-step:

1. Electricity Supply: When direct current (DC) power is applied, typically sourced from renewable energy or grid power, it provides the energy needed to drive the chemical reaction in the electrolyser. The anode side of the cell, which is positively charged, attracts electrons to initiate the reaction.
2. Water Molecule Splitting: Water enters the cell on the anode side, where the electrical current splits each water molecule into oxygen, protons (H^+ ions), and electrons. This reaction produces pure oxygen at the anode surface as a byproduct.
3. Proton Movement Through the Membrane: The PEM allows the H^+ ions (protons) to travel through the membrane to the cathode side of the cell. This selective permeability of the electrolyte membrane is critical because it prevents oxygen and hydrogen gases from mixing, reducing the risk of explosions and enhancing purity.
4. Hydrogen Production at the Cathode: On the cathode side, electrons coming out from the anode combine with the protons that have passed through the membrane, producing atomic hydrogen, and then molecular hydrogen. This hydrogen is collected and stored for various applications, from industrial uses to clean energy generation. Hydrogen is evolved at the cathode, and oxygen at the anode.

5. Gas Collection and Purification: Both oxygen and hydrogen gases are separately collected. Hydrogen produced by PEM electrolysis is very pure (typically over 99.999%), which is ideal for applications requiring high-grade hydrogen, such as fuel cells.

The solid polymer electrolyte membrane is a key advantage of this process, enabling the electrolysers to start and stop quickly, respond well to fluctuations in electricity, and produce high-purity hydrogen.

Pros of PEM Electrolysers

- High Efficiency: PEM electrolysers achieve high energy efficiency, making them preferred over others. The energy efficiency of this type of electrolysers is 80%, with a HHV of hydrogen, as claimed.
- Fast Response Time: Their rapid startup and load adjustment capabilities make them well-suited for use with intermittent renewable energy sources like wind and solar.
- Compact and Lightweight Design: Small size allows for installation in limited spaces, which can be advantageous for certain projects.
- High-Purity Hydrogen Production: PEM technology produces high-purity hydrogen, 99.999% plus, minimising the need for additional purification.
- Lower Operational Temperatures: Lower temperature reduces the degradation rate of materials, potentially extending equipment lifespan.

Cons of PEM Electrolysers

- While PEM electrolysers offer some benefits, they also come with significant limitations that need to be carefully considered:
- High Initial Cost: The costly materials needed, including precious metal catalysts (platinum and iridium) and polymer electrolytes. This drives up the cost of PEM electrolysers, making them a larger investment than alkaline systems.
- Limited Lifespan: Despite lower operational temperatures, PEM membranes and electrodes are more prone to wear and tear over time. This can result in a shorter operational life compared to alkaline electrolysers.
- Dependency on Precious Metals: PEM electrolysers rely on platinum and iridium as catalysts, which are both costly and subject to market fluctuations and availability as well. This can make scaling up PEM production financially challenging. Availability of critical metals, including rare earth elements, is limited due to complex mining and extraction.

- Membrane Degradation Under Stress: While PEM electrolysers respond quickly to variable loads, the constant adjustment can accelerate membrane degradation, impacting lifespan and overall cost-effectiveness.

5.6.3 Anion Exchange Membrane (AEM) Electrolysis:

AEM electrolysis combines features of both alkaline and PEM technologies. It uses an anion-conducting membrane and can operate with both alkaline and non-alkaline electrolytes. AEM is considered a promising technology for the future due to its potential for high efficiency and lower cost.

- Pros: No need for precious metals as catalysts, making them potentially more affordable than PEM electrolysers.
- Cons: AEM technology is still in the early stages of development and lacks commercial maturity and reliability.

5.6.4 Solid Oxide Electrolysis (SOE):

A solid oxide electrolyser (SOEC) is a high-temperature (typically 600 to 900°C) electrochemical device that uses a solid ceramic electrolyte to split water. High-temperature operation makes it more efficient than others because it uses external heat, reducing the electrical energy required. SOECs can also run in reverse as fuel cells, a unique ability that allows them to convert hydrogen and oxygen back into electricity when needed. It has yet to be fully commercialised.

- Pros: Superior efficiency thanks to heat reuse, making them ideal for industrial applications with access to excess heat.
- Cons: The high temperatures involved lead to faster material degradation, increased maintenance needs, and reduced lifespan.

5.6.5 Electrolyser Capacity

The capacity of an electrolyser is expressed in terms of plug-in capacity from the power grid. If the plug-in capacity of an electrolyser is 1MW, it means the electrolyser can be plugged into a 1MW power supply; however, H_2 production will depend on the electrolyser's energy efficiency, which ranges from 65–70%.

The energy efficiency of an electrolyser is defined as the heat energy produced in the form of hydrogen, based on the HHV of the hydrogen, divided by the electric energy consumed. While theoretically, electrolysis could be highly efficient, real-world systems typically range from 70 to 75% energy-efficient, on the basis of the HHV of hydrogen. Alkaline and PEM electrolysers are common types that are used by industries. PEM is more efficient than the alkaline type.

Voltage efficiency in an electrolyser refers to how effectively it converts electrical energy into chemical energy in the form of hydrogen, specifically by comparing the voltage required to the thermoneutral voltage (Vtn) of the water dissociation reaction.

Voltage efficiency (η) is defined as the ratio of the thermoneutral voltage (Vtn) to the real voltage (V) applied to the electrolysis cell, η =Vtn / V.

It is a crucial metric for evaluating the performance of electrolysers. A higher voltage efficiency indicates that the electrolyser is operating closer to the ideal, thermally neutral condition (when no external heat is required), requiring less energy to produce hydrogen. There is a loss of voltage due to activation polarisation, concentration polarisation, resistance to the flow of ions (ohmic loss), and bubble formation at the electrode. Overcoming all these resistive factors reduces the supplied voltage actually used for splitting water molecules, thereby reducing the electrolyser efficiency.

Electrolyser Rating - Electrolysers are rated by their electrical power consumption to produce 1 kg of hydrogen per hour. The power rating and energy efficiency of an electrolyser determine how much hydrogen it can produce in a given time.

An electrolyser with a capacity of 1 MW and 70% energy efficiency can produce (1,000 x 24 x 0.7)/39.4 = 426.39 kg/day, or 17.76 kg/hour, of hydrogen. The HHV of hydrogen (39.4 kWh/Kg) has been considered.

6 Green Hydrogen

Green hydrogen is produced from water electrolysers, fed with green electric energy derived from solar or wind.

There are 2 main types of electrolysers, which are mostly used for green hydrogen production: alkaline and proton exchange membrane (PEM).

A process of hydrogen production from an electrolyser consists of 3 units: a water treatment unit, an electrolysis unit, and a hydrogen purification unit. The water treatment unit makes demineralised water suitable for electrolysis. The electrolysis unit consists mainly of an electrolyser with a DC power supply arrangement, and the purification unit consists of a hydrogen separator, a wash unit, compression, and storage.

The process flow diagrams of a typical alkaline and PEM electrolyser are shown below in Figures18 and 19.

6.1 Hydrogen from Alkaline Electrolyser.

The alkaline electrolyser has been used to make hydrogen for more than a hundred years. Early in the twentieth century, a little over 400 industrial electrolysers were operating in Europe, driven by hydropower-generated electricity, producing hydrogen for fertiliser to ensure Europe's wheat supply. The biggest containerised alkaline electrolysers are being built by Nel in Norway. These can be stacked, one against the other, and wired together. Glomfjord, Norway, which operated from 1953 to 1991, had a staggering capacity of as high as 135 MW.

Alkaline electrolysers are relatively cheap because their component materials are cheap and easily accessible, and are getting cheaper still. This uses an alkaline solution (usually potassium hydroxide or sodium hydroxide) as the electrolyte. The electrodes of an alkaline electrolyser are commonly made of stainless steel (SS 304, SS 316L) or Cu-Ni alloy. However, the hydrogen produced by alkaline electrolysers has a lower purity of 99.98%, with corrosive gases and water vapour as thc main impurities.

They are reliable and a good way of making a stable supply of hydrogen if there is a stable supply of electric powcr. They are, however, bulky, and the liquid inside them is often corrosive. It takes time to ramp them up and down.

A process flow diagram of an alkaline electrolyser is given in Figure 18 below.

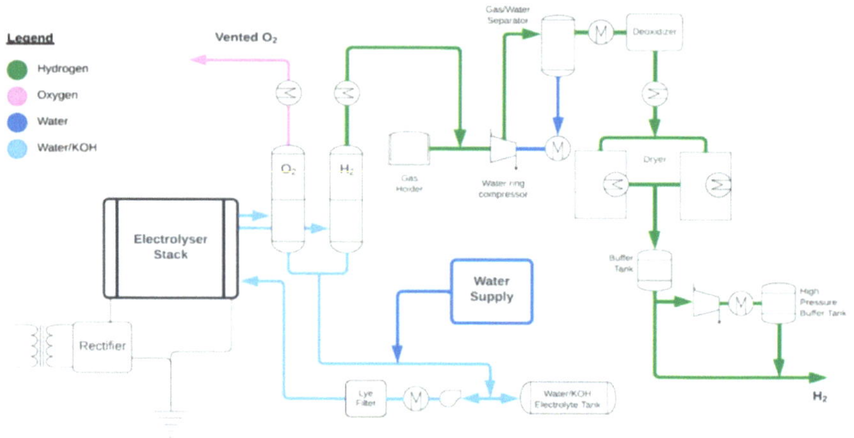

▲ Figure 18: Typical Process flow Diagram for an Alkaline Electrolyzer (Note: O_2 = oxygen; H2 = hydrogen; KOH = potassium hydroxide)

6.2 Hydrogen from PEM Electrolyser

There is a new, advanced electrolyser: a proton exchange membrane (PEM). It was conceived in the mid-1960s by General Electric to produce electricity for NASA's space programme, which was brought to the commercial market by ABB in 1987. In a PEM electrolyser, protons cross the membrane from the anode to the cathode. Proton (H^+) passes through a solid electrolyte, a specially built polymer that blocks the negatively charged ions. H^+ picks up an electron, coming out from the anode, at the cathode to become atomic hydrogen and then molecular hydrogen. Hydrogen purity is as high as 99.999%. This arrangement does away with the bulky liquid electrolyser.

PEM electrolyser starts up more quickly than alkaline electrolyser, making it ideal for harvesting intermittent energy sources like solar and wind. It is smaller than the equivalent alkaline electrolyser, stackable, easier to maintain, more reliable, and simpler to operate. It is efficient, converting around75–80% of electrical energy to chemical energy in the form of hydrogen, considering the higher heating value (HHV) of hydrogen.

There are a few downsides to PEM. After about 40,000 hours of continuous operation, the electrolyte membrane degrades beyond use. It is costly because of the costly electrodes (costly PGMs) and electrolytic polymer.

Siemens has been operating a 6 MW PEM electrolyser since 2015 in Mainz, Germany since 2015. ITM Power Ltd has secured a joint venture project with Shell to construct a 10 MW PEM electrolyser in the Rhineland Refinery Complex in Germany. REFHYNE, the largest PEM electrolysis plant in the world, is now operational. REFHYNE is an acronym for Clean Refinery Hydrogen for Europe. REFHYNE is a project focused on developing and implementing large-scale renewable hydrogen production within an industrial setting to reduce carbon emissions. It will use renewable energy to produce about 4 tonnes of clean hydrogen per day.

Most likely, due to their rapid response capabilities, PEM electrolyser will be widely used to manage intermittent renewable energy sources, whereas alkaline will be used in more industrial applications, where cost and reliability are determining factors.

A process flow diagram of the PEM electrolyser is shown in Figure 19.

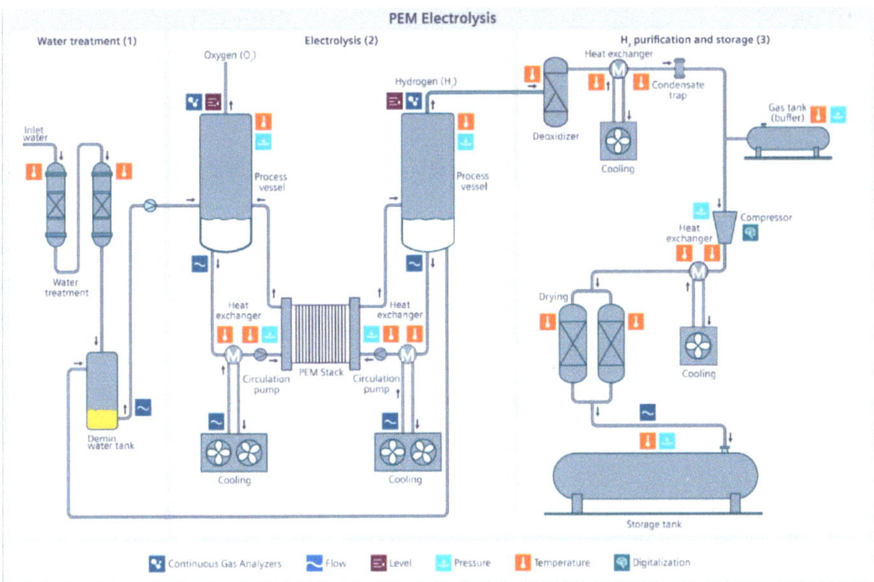

▲ Figure 19: Example of a PEM (Porton Exchange Membrane) Electrolyzer Process Chart

6.3 Comparison Between PEM and Alkaline Electrolyser.

A comparison of alkaline and PEM analysers, with respect to their technical and operational capabilities are given below in Tables 6 and 7.

PEM Electrolysers vs. Alkaline Electrolysers

▼ Table 6: Qualitative comparison of alkaline and PEM electrolysers.

Feature	Alkaline	PEM
Cost	Lower	Higher
Efficiency	Good	Better
Startup Time	Longer	Fast
Response to Load Changes	Slower	Rapid
Hydrogen Purity	Good	Better
Durability	High	Moderate
Scalability	Suitable for large-scale	Limited
Operating Temperature	60–80°C	50–70°C
Maintenance Complexity	Simple	Complex
Material Cost	Lower	Higher

▼ Table 7: PEM Hydrogen Generator vs Alkaline Hydrogen Generator

Parameter	Alkaline Electrolyzer	PEM Electrolyzer	SOEC Electrolyzer	AEM Electrolyzer
Electrolyte	20% ~ 30% KOH	Perfluorosulfonic Membrane	Yttria-Stabilized Zirconia	Divinylbenzene Polymer Carrier
Diaphragm	Asbestos/Polyphenylsulfone, etc.	Perfluorosulfonic Membrane	Zirconia-based Ceramic Membrane	Anion Exchange Membrane
Anode Catalyst	Nickel-plated Perforated Stainless Steel	Iridium Oxide	Perovskite-type	High Surface Area Nickel or NiFeCo Alloy
Cathode Catalyst	Nickel-plated Perforated Stainless Steel	Platinum Nanoparticles	Nickel/Zirconia	High Surface Area Nickel
Operating Temperature/°C	70 ~ 90	50 ~ 80	700 ~ 900	40 ~ 60
Working Pressure/MPa	0.1 ~ 3.0	4.0 ~ 7.0	0.1	<3.5
System Efficiency/%	60 ~ 75	70 ~ 90	85 ~ 100	60 ~ 90
Hydrogen Purity	≥99.8%	≥99.99%	≥99.99%	≥99.99%
Current Density (A-cm-2)	0.2 ~ 0.6	1.0 ~ 2.0	1.0 ~ 10	0.2 ~ 0.4
Energy Consumption (kW-h-m-3)	4.2 ~ 5.9	<4.3	>3.7	——
Hydrogen Pressure (MPa)	0.1 ~ 1.0	>2.0	>0.01	——
Lifespan (h)	6000-80000	8000-60000	20000	/
Porous Transport Layer Anode	Nickel Mesh	Platinum-Plated Sintered Porous Titanium	Nickel Mesh or Foam Nickel	Foam Nickel
Porous Transport Layer Cathode	Nickel Mesh	Sintered Porous Titanium or Carbon Cloth	None	Foam Nickel or Carbon Cloth
Bipolar Plate Anode	Nickel-plated Stainless Steel	Platinum-Plated Titanium	None	Nickel-plated Stainless Steel
Bipolar Plate Cathode	Nickel-plated Stainless Steel	Gold-Plated Titanium	Cobalt-plated Stainless Steel	Nickel-plated Stainless Steel
Technical Maturity Level	8 ~ 9	8 ~ 9	5 ~ 6	2 ~ 3
Industrialization Level	Fully Industrialized	Fully Industrialized	Laboratory to Industrial Transition	Laboratory to Industrial Transition
Advantages	Simple, Technologically Mature, High Reliability, Can Operate at Normal Temperature and Pressure	Long Lifecycle, Good Stability, Low Electrolyzer Corrosion, High Electrolysis Efficiency, Simplified System, Compact Device Structure, High Hydrogen Purity	High Electrolysis Efficiency, Can Reach More than 90%, Low Energy Consumption, Low Cost	Combines the Advantages of Alkaline Water Electrolysis and Proton Exchange Membrane Water Electrolysis
Disadvantages	Low Hydrogen Production Efficiency, High Energy Consumption, Alkali Leakage Environmental Pollution Issues	High Cost ; Requires Precious Metal Catalysts	High Operating Temperature of 600 ~ 1000°C, High Temperature Requirements, Key Materials Easily Age at High Temperatures	Polymer Membrane has Low Hydroxide Ion Conduction Rate, Poor Stability; Catalyst Easily Poisoned by Metal Ions

The values in the table above are estimations based on the average specifications of each technology.

6.4 Global Deployment of Water Electrolyser

The deployment of electrolysis capacity (which is actively operational) grew by around 60%, reaching 1.75 GW as of May 24 (Hydrogen Insight 24 by McKinsey & Company), up from 1.1 GW in 2023. The global installed capacity of green hydrogen equals about 185 kt per annum. Green hydrogen accounts for about 20% of installed clean hydrogen production capacity globally. Green and blue hydrogen together make clean hydrogen.

Global installed capacity of water electrolysis reached 2 gigawatts (GW) in 2024. China now accounts for 65% of global installed capacity and capacity that has reached a final investment decision. China is also home to nearly 60% of global electrolyser manufacturing capacity. China's existing manufacturing capacity of 20 GW per year is significantly above current demand, which was around 2 GW in 2024. The cost of making and installing an electrolyser outside China in 2024 was USD 2,000 to USD 2,600 per kilowatt (kW), compared with USD 600 to 1,200 per kW for electrolysers manufactured and installed in China, according to the IEA 2025 report.

Today, the largest deployed electrolyser capacity is in China (1,150 MW), where the world's 2 largest operational projects with capacities of 260 MW and 150 MW are located. In China, 95% of deployed electrolysis capacity is based on alkaline technology, whereas PEM technology is more prevalent in Europe and North America, accounting for 60% of total installed capacity.

Figure 19 shows the volume and types of electrolysers installed from 2019 to May 2024 by different countries.

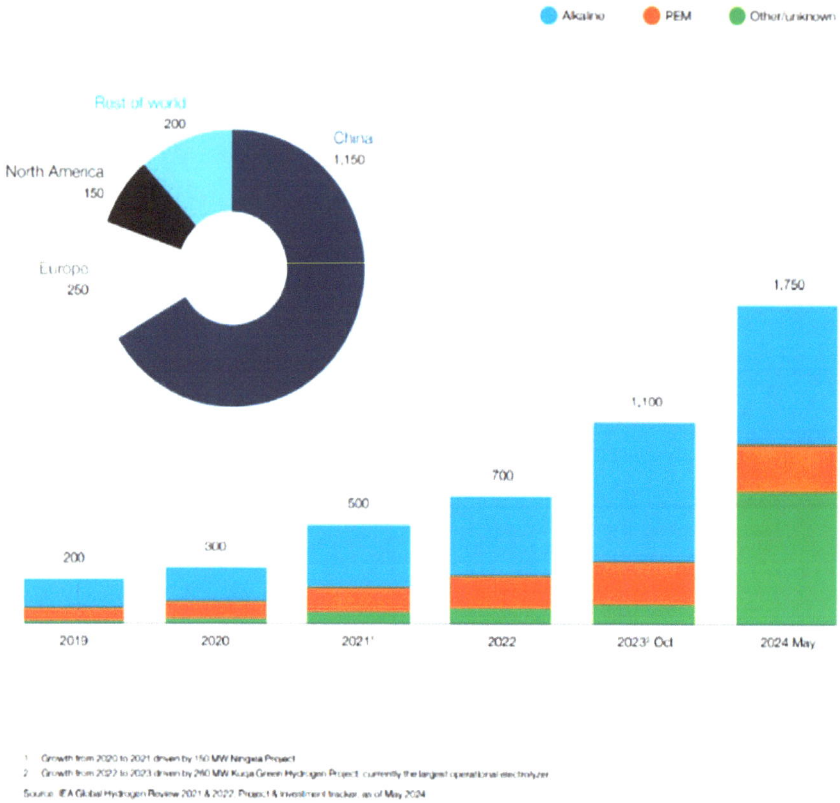

▲ Figure 20: Global Cumulative Installed Electrolysis Capacity, MW

Technological Advancement in Electrolysers

Technological advancements in water electrolysis focus on boosting efficiency, lowering costs, improving integration with renewables, better catalysts, advanced membranes, and optimised stack designs, allowing the use of impure water and unlocking scalable green hydrogen production for decarbonisation. Key progress includes high-temperature SOECs, semi-vapour electrolysis for seawater desalination, and improved integration with variable renewable energy sources.

Key technological advancements in different areas are as follows:

- Material & Catalyst Innovation: Developing cheaper, more active catalysts (reducing reliance on PGMs) and robust, selective membranes for better ion conduction and gas separation.

- System Efficiency & Integration:
  - PEM (Proton Exchange Membrane): Higher efficiency, fast response for renewable integration, compact design.
  - SOEC (Solid Oxide Electrolyser Cells): High-temperature operation (700–1,000°C) offers >85% efficiency but requires longer startup; advanced designs are improving this.
  - AEM (Anion Exchange Membrane): Combines the benefits of alkaline (lower cost) and PEM (membrane) systems, reducing membrane lifetime issues in PEM.
- Water Source Flexibility: New methods, like semi-vapour electrolysis, allow the use of unpurified water sources (e.g., seawater) by avoiding ion impact, significantly reducing purification costs.
- Nanotechnology: Deep-sub-Debye-length nanogap cells (typically <100 nm) enable efficient electrolysis of pure water by overlapping electrical double layers, enhancing ion transport. By creating a strong field across the gap, they enhance ionisation and mass transport, bypassing the need for conductive ions. This "virtual breakdown mechanism" enables higher current densities for cleaner hydrogen production.
- System Architecture: Improved cell/stack design and integration with renewable energy sources (solar, wind) to manage fluctuating power input and optimise energy use.

Impact & Future Focus

- Cost Reduction: Advancements aim to drastically cut the cost of green hydrogen, making it competitive with fossil fuels.
- Scalability: Addressing challenges in scaling up electrolyser sizes for industrial-level production.
- Decarbonisation: Enabling large-scale green hydrogen production, a key component in achieving global zero-carbon goals.

The advancement targets for alkaline, PEM, and solid oxide electrolysers by 2050, taken from the ResearchGate journal, are given in Tables 7a, 7b and 7c, respectively.

▼ Table 7a: Target of Alkaline Electrolyser

Parameter	2020	Target 2050	R&D focus
Nominal current density	0.2–0.8 A cm^{-2}	> 2 A cm^{-2}	Diaphragm
Voltage range (limits)	1.4–3 V	< 1.7 V	Catalysts
Operating temperature	70–90 °C	> 90 °C	Diaphragm, frames, balance of plant components
Cell pressure	< 30 bar	> 70 bar	Diaphragm, cell, frames
Load range	15–100 %	5–300 %	Diaphragm
Hydrogen purity	99.9–99.9998 %	> 99.9999 %	Diaphragm
Voltage efficiency (LHV)	50–68 %	> 70 %	Catalysts, temperature
Electrical efficiency (stack)	47–66 kWh kg^{-1} H_2	< 42 kWh kg^{-1} H_2	Diaphragm, catalysts
Electrical efficiency (system)	50–78 kWh kg^{-1} H_2	< 45 kWh kg^{-1} H_2	Balance of plant
Lifetime (stack)	60,000 hours	100,000 hours	Electrodes
Stack unit size	1 MW	10 MW	Electrodes
Electrode area	10,000–30,000 cm^2	30,000 cm^2	Electrodes
Cold start (to nominal load)	< 50 minutes	< 30 minutes	Insulation (design)
Capital costs (stack) minimum 1 MW	USD 270/kW	< USD 100/kW	Electrodes
Capital costs (system) minimum 10 MW	USD 500–1000/kW	< USD 200/kW	Balance of plant

▼ Table 7b: Target of PEM Electrolyser

Parameter	2020	Target 2050	R&D focus
Nominal current density	1–2 A cm^{-2}	4–6 A cm^{-2}	Design, membrane
Voltage range (limits)	1.4–2.5 V	< 1.7 V	Catalyst, membrane
Operating temperature	50–80 °C	80 °C	Effect on durability
Cell pressure	< 30 bar	> 70 bar	Membrane, reconversion catalysts
Load range	5–120 %	5–300 %	Membrane
Hydrogen purity	99.9–99.9999 %	Same	Membrane
Voltage efficiency (LHV)	50–68 %	> 80 %	Catalysts
Electrical efficiency (stack)	47–66 kWh kg^{-1} H_2	< 42 kWh kg^{-1} H_2	Catalysts/membrane
Electrical efficiency (system)	50–83 kWh kg^{-1} H_2	< 45 kWh kg^{-1} H_2	Balance of plant
Lifetime (stack)	50,000–80,000 hours	100,000–120,000 hours	Membrane, catalysts, PTLs
Stack unit size	1 MW	10 MW	MEA, PTL
Electrode area	1500 cm^2	> 10,000 cm^2	MEA, PTL
Cold start (to nominal load)	< 20 minutes	< 5 minutes	Insulation (design)
Capital costs (stack) minimum 1 MW	USD 400/kW	< USD 100/kW	MEA, PTLs, BPs
Capital costs (system) minimum 10 MW	USD 700–1400/kW	< USD 200/kW	Rectifier, water purification

▼ Table 7c: Target of Solid Oxide Electrolyser

Parameter	2020	Target 2050	R&D focus
Nominal current density	0.3–1 A cm^{-2}	> 2 cm^{-2}	Electrolyte, electrodes
Voltage range (limits)	1.0–1.5 V	< 1.48 V	Catalysts
Operating temperature	700–850 °C	< 600 °C	Electrolyte
Cell pressure	1 bar	> 20 bar	Electrolyte, electrodes
Load range	30–125 %	0–200 %	Electrolyte, electrodes
Hydrogen purity	99.9 %	> 99.9999 %	Electrolyte, electrodes
Voltage efficiency (LHV)	75–85 %	> 85 %	Catalysts
Electrical efficiency (stack)	35–50 kWh kg^{-1} H_2	< 35 kWh kg^{-1} H_2	Electrolyte, electrodes
Electrical efficiency (system)	40–50 kWh kg^{-1} H_2	< 40 kWh kg^{-1} H_2	Balance of plant
Lifetime (stack)	< 20,000 hours	80,000 hours	All
Stack unit size	5 kW	200 kW	All
Electrode area	200 cm^2	500 cm^2	All
Cold start (to nominal load)	> 600 minutes	< 300 minutes	Insulation (design)
Capital costs (stack) minimum 1 MW	> USD 2000/kW	< USD 200/kW	Electrolyte, electrodes
Capital costs (system) minimum 10 MW	Unknown	< USD 300/kW	All

It is worth mentioning here that Japan has developed an underwater turbine designed to generate continuous renewable energy from powerful ocean currents. The system, created by IHI Corporation, is anchored to the seabed and uses massive blades to spin with the natural flow of water. Unlike wind or solar power, ocean currents are constant at the seabed, providing a stable and predictable source of clean energy. This breakthrough highlights Japan's efforts to reduce dependence on fossil fuels and expand sustainable energy options. Research shows that ocean current energy could one day supply a significant portion of the electricity needs of coastal nations.

Hydrogen, transported through a pipeline, can transport energy 10 times more than electric transmission at one-eighth the cost of a power transmission line. Turbine power can be converted into hydrogen molecules nearby and then transported to shore. This green power can be used to produce green hydrogen, which can be transported and distributed like natural gas. Scientists see it as a reliable alternative to intermittent renewable sources. Figure 21 shows an underwater turbine.

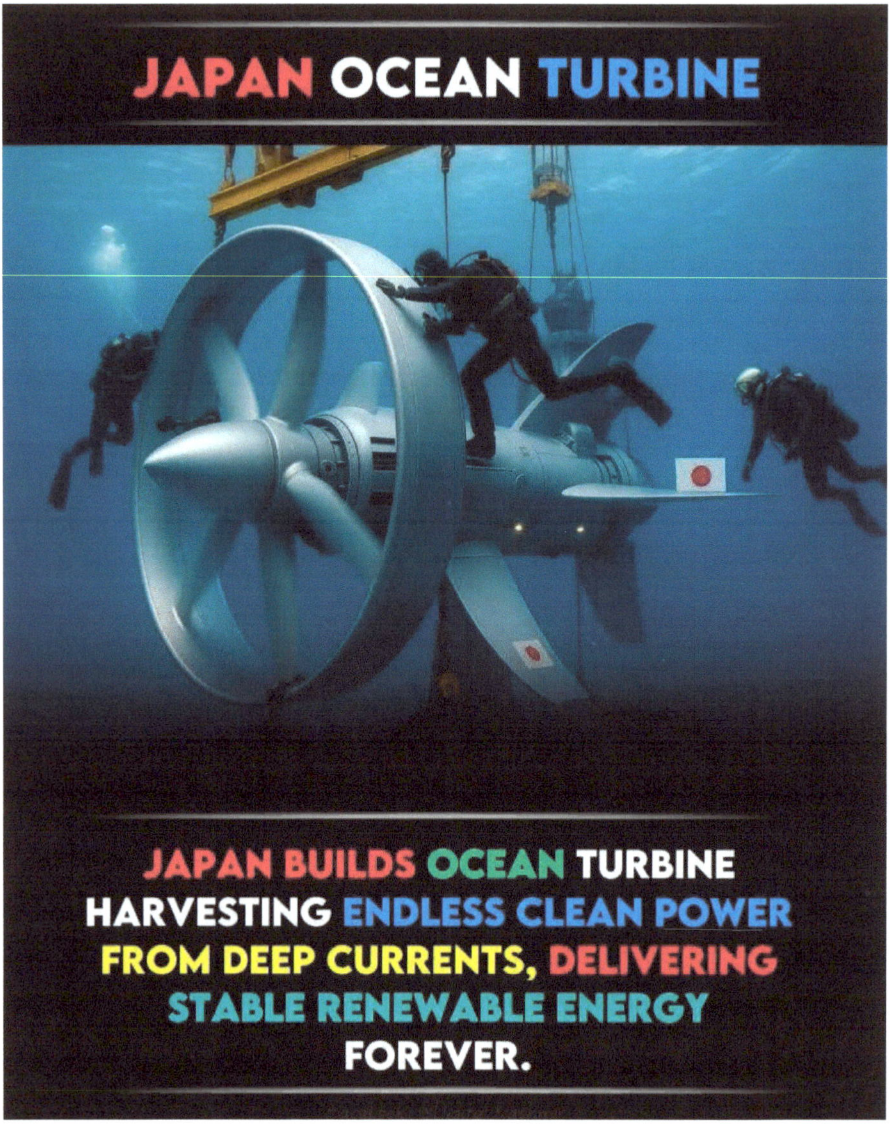

▲ Figure 21: Underwater Power Turbine

6.5 Electrolysers and Critical Metals

Critical metals for electrolysis, for green hydrogen production, include platinum, iridium, nickel, and REEs such as lanthanum and yttrium. These metals are essential for the catalysts and electrolytes used in electrolysers.

The demand for these critical metals is projected to increase significantly as the world scales up green hydrogen production to meet climate goals. This has led to concerns about potential supply shortages and price volatility, especially for platinum, iridium, and some other REEs. Research efforts are focused on developing alternative materials and optimising electrolyser designs to reduce reliance on these critical metals.

Mining of critical metal ores and their extraction to the required purity are environmentally sensitive activities, given excavation, deforestation, and water requirements. The extraction processes, like solvent extraction, ion exchange, electrowinning, and metallothermic, require huge amounts of water and energy. The extraction technology is advanced because ores contain very low and ultralow concentrations of metals. This technology is mostly centred in China.

Critical metal ores are found in selected countries such as China, South Africa, Chile, Australia, and Indonesia. The trijunction of Chile, Argentina, and Bolivia, known as the lithium triangle, is reported to have large lithium reserves in saline water. Lithium batteries are essential for energy storage to ensure un interrupted supply of power to electrolysers. Extraction from saline water is less environmentally sensitive.

6.6 Critical Metal Demand and Production

To produce green hydrogen from electrolysers, critical minerals are required at 3 levels of the supply chain: at the level of electricity production from solar and wind, at the level of electrolytic green hydrogen production, and finally at the level of energy storage in batteries.

By 2050, global critical metal requirements are projected to increase significantly, with some metals seeing 8 to 17-fold increases from 2023 levels, driven by demand for clean energy technologies such as electric vehicles and renewable energy sources. The IEA estimates that a net-zero emissions scenario could require a 6 billion tonne increase in metals production from 2024 to 2050. This growth necessitates substantial investments in mining, extraction, and critical metal recycling to meet anticipated demand and avoid potential supply shortages.

The Breakthrough Institute has developed a specific model to predict the mineral requirements. In this analysis, they have used future electrolyser and fuel cell capacities to derive material requirements. Modelling of global electrolyser capacity in the 2050 range widely varies between 1,400 and 7,800 GW. The scenarios used for this analysis are the IEA's SDS (sustainable development scenario), IRENA, and the Economic Transitions Commission (ETC).

Electrolyser capacities for the years 2030, 2040, and 2050 are shown in Table 8. The cumulative material requirement for the years 2030, 2040, and 2050 is shown in Table 9.

Market shares of alkaline, PEM, and solid oxide electrolysers have been assumed to be 40, 40, and 20%, respectively, in their analysis when determining the requirement.

▼ Table 8: Future Modelled- Electrolyser Capacity

Future Modeled Electrolyzer Capacity in 2030, 2040, and 2050

	IEA SDS	IRENA	ETC
2030 Cumulative Electrolyzer capacity (GW)	150	270	225
2040 Cumulative Electrolyzer capacity (GW)	500	2,885	3,300
2050 Cumulative Electrolyzer capacity (GW)	1,400	5,500	7,800

Get the data · Created with Datawrapper

Projected global electrolyser capacity in 2030, 2040, and 2050 in three selected future world hydrogen market analyses. The IRENA 2040 Cumulative Electrolyser Capacity was approximated by assuming a linear relationship between the 2030 and 2050 capacities.

▼ Table 9: Cumulative material requirement for electrolysers

Cumulative Material Requirement in 2030, 2040, and 2050

		2030	2040	2050	2030	2040	2050
	Cumulative Mineral Requirement (tons)	IEA SDS	IEA SDS	IEA SDS	IRENA	IRENA	IRENA
Alkaline	Nickel	48,000	160,000	448,000	86,400	923,200	1,760,000
	Zirconium	6,000	20,000	56,000	10,800	115,400	220,000
	Aluminum	30,000	100,000	280,000	54,000	577,000	1,100,000
PEM	Platinum	18	60	168	32	346	660
	Palladium	18	60	168	32	346	660
	Iridium	42	140	392	76	808	1,540
SOEC	Nickel	5,250	17,500	49,000	9,450	100,975	192,500
	Zirconium	1,200	4,000	11,200	2,160	23,080	44,000
	Lanthanum	600	2,000	5,600	1,080	11,540	22,000
	Yttrium	150	500	1,400	270	2,885	5,500
Fuel Cells	Platinum	145	2,320	5,800			

Get the data • Created with Datawrapper

Cumulative Modal Electrolyzer and Fuel Cell Material Requirements 2030-2050 using 40%-40%-20% market share for Alkaline-PEM-SOEC Electrolyzers. The IEA SDS predicts the lowest material demands, while ETC's scenario anticipates the highest requirements. Please note that the Aluminum requirement here refers to Aluminum metal.

Material requirements for electrolysis vary by cell architecture. There are 4 generations of electrolysis technology to produce low-carbon hydrogen. These are alkaline, PEM, solid oxide, and AEM. The material requirements of these types are given in Table 9.

▼ Table 10: Material requirement for electrolyser of different architecture

The material requirement for electrolysis technology will vary by cell architecture

	AEC	PEMEC	SOEC	AEMEC
Electrolyte	Aqueous potassium hydroxide	Solid electrolyte	Solid oxide electrolyte	Aqueous potassium hydroxide
Separator/ membrane	Zirconium oxide	Fluorinated polymer membranes (Nafion)	Yttria-stabilized zirconium (YSZ)	Aromatic and aliphatic hydrocarbons polymers
Anode catalyst (oxygen side)	Nickel alloy-plated steel/nickel-cobalt	Ruthenium or Iridium oxides/platinum-carbon	Nickel/YSZ	Nonplatinum group metals/nickel-based alloy
Cathode catalyst (hydrogen side)	Nickle-based alloys	Platinum or platinum-palladium	Lanthanum-strontium-manganite/YSZ	Nickel-iron-molybdenum grown on nickel foam
Bipolar plates	Nickel-coated stainless steel	Platinum/gold-coated titanium	Stainless steel/iron-nickel alloy	Nickel-coated stainless steel
Frames and sealing	Sulfonate/ fluorinated polymers	Sulfonate/fluorinated polymers	Ceramics/glass	Silicon/fluorinated polymers

Supply and criticality risk of key minerals for hydrogen production have been drawn in the analysis report by Lux Research Gate, which is shown in Figure 22.

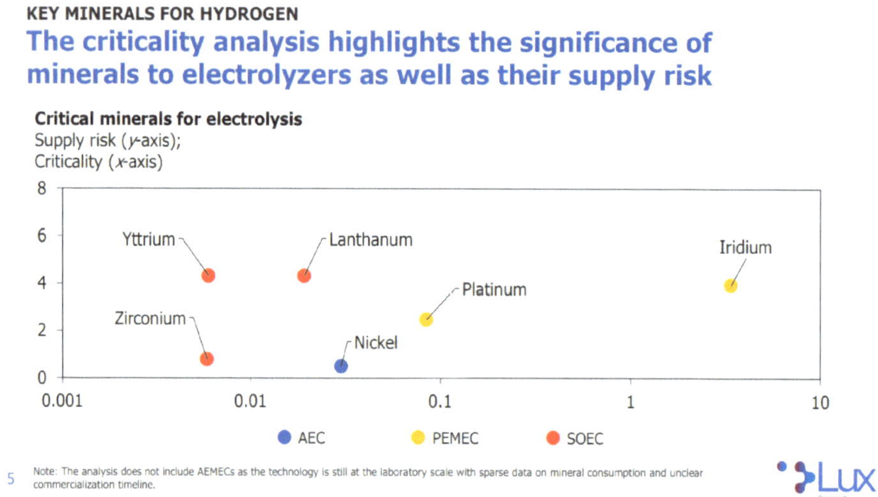

▲ Figure 22: Supply risks and criticality of different metals in electrolysers.

Nickel – Nickel is widely used across water electrolysis technologies for hydrogen production. Today, a 1-GW electrolyser uses anywhere between 800 and 1,000 tonnes of nickel. From the chart in the previous figure, nickel has a lower criticality score (less than 1) than other minerals. It also has a low supply risk, since as many as 25 countries around the world have nickel reserves. However, its widespread use in other energy applications, such as fast-growing battery energy storage, has led nickel prices to soar to a decade-high. The IEA estimates that a further increase in nickel prices could have significant effects on future battery and electrolyser supply chains.

Iridium – Iridium is far scarcer than platinum. However, due to its excellent catalytic activity in highly acidic conditions, finding promising alternatives is a challenge. Analysis also shows that while platinum is a crucial mineral for PEM electrolysers to flourish in the future, the world is unlikely to run out of it. However, supply risk is relatively high for both minerals as PGM production is concentrated in South Africa.

Zirconium, yttrium, and lanthanum – While zirconium is a non-critical mineral, yttrium and lanthanum show up in the high supply risk region of the chart. China produces nearly 4 times as much rare earths as the second-largest producer, the US. Nearly 95% of these mineral reserves are in China, increasing their supply risk, as solid oxide electrolyser technology scales up to decarbonise industrial applications. A lack of strong alternatives to these minerals adds to the criticality of yttrium and lanthanum.

Lux Research Gate, in its publication *Critical Minerals in the hydrogen economy*, finds that iridium is the biggest bottleneck to the multi-GW installation of water electrolysis plants, whereas other crucial PGMs. and rare earths, although abundant, are strongly concentrated in a few countries, threatening the future rollout of electrolysis technologies. Clients should consider investing in finding alternative materials for these critical minerals used in electrocatalysts and investing early in a recycling infrastructure, which would alleviate some of these choke points in future supply.

While electrolysers available today are ready for commercial applications, there is still huge potential to innovate. New innovations in zero-gap electrolysis or membrane-less electrolysers could unlock at least 10% to 15% efficiency improvements that can result in lower capex installations. Clients should monitor developments in emerging electrolyser technologies that will potentially displace existing electrolysers.

6.7 Recycling of Critical Minerals

Most electrolyser stacks typically last 7 to 10 years before refurbishment or replacement. As electrolyser companies focus on GW-scale manufacturing facilities, they are expected to partner with EPC firms to not only optimise operations and maintenance and refurbish stacks, but also, more importantly, enable the recycling of stacks in partnership with metals recyclers. There is expected to exist an electrolyser recycling industry. Companies should look to supply chain partnerships like opportunities we see in the battery recycling market today, where recyclers are partnering across the battery production value chain. In the electrolyser market, this could mean partnerships between mining players, stack producers, EPCs, and recyclers. There are initiatives to fund sustainable hydrogen manufacturing and electrolyser recycling in the US and Europe. By 2035, clients should expect to see complete electrolysers being recycled, as technologies like automated sorting, dismantling, metal extraction, and purification become mainstream.

6.8 Critical Minerals and Global Electrification.

Electrification of the global energy system is spurring renewed efforts into a broader shift from thermochemistry to electrochemistry. This shift has resulted in new developments, such as CO_2 electrolysers for producing hydrocarbon chemicals, direct ammonia synthesis through electrolysers, molten-oxide electrolysis for refining iron ore, and even electrolysers for producing concrete from electricity. The development and eventual deployment of these novel electrochemical platforms will disrupt the demand for critical metals and REEs, which are rare and in short supply. It is therefore important to monitor the innovation landscape of electrochemistry and pre-empt any material supply challenges that may arise in the way of decarbonising the planet.

IRENA has raised concerns about the global supply of raw materials critical to the production of some electrolysers, including platinum and iridium, arguing that increased demand could drive up prices and thus raise the cost of hydrogen-producing machines.

The IRENA report says that it expects the demand for electrolysers to increase to 5,000 GW by 2050 in a 1.5°C scenario.

The alkaline electrolysers that dominate the market today rely on materials generally considered non-critical, such as steel and nickel. By contrast, polymer electrolyte membrane (PEM) electrolysers and solid oxide electrolysers (SOE) appear to pose more serious problems of critical material dependence.

Platinum and iridium, used in polymer electrolyte membrane electrolysers as catalysts, are the scarcest and emissions-intensive metals in the world. Their supply is also highly concentrated, with South Africa supplying over 70% of global platinum and over 85% of global iridium. Currently, no substitutes for iridium are available or foreseen. The current iridium supply would only support annual PEM electrolyser deployment of 3-7.5GW as reported.

This could prove to be problematic, given that manufacturers including ITM Power, Cummins, Plug Power, and Ohmium have all announced plans to scale up their annual PEM electrolyser production by at least 11GW between them.

The price of iridium more than tripled due to supply-bottleneck concerns, rising from just under $70,500/kg in December 2020 to a high of $ 222,075/kg in May 2021, after producer Anglo American Platinum (Amplats) closed a processing facility in South Africa for several months, as reported.

As more countries pursue deep decarbonisation strategies, hydrogen will play a critical role in their energy transition. Precious metals and minerals are deeply embedded in renewables and electrification technology. The hydrogen economy is no different. Key minerals such as platinum, iridium, nickel, and REEs play a crucial role in the rollout of hydrogen production through electrolysis.

India imports most of the critical metals, such as lithium, cobalt, nickel, platinum, iridium, and REEs, needed for electrolysers and other advanced technologies. This is likely to be one of the biggest constraints due to changing global geopolitical conditions.

India is entering into strategic partnerships with other nations and plans to lease mines in Africa and Latin America. Critical metal security has become very important for India, which is developing fast and has a vision to join the club of developed nations by 2047. Encouragement and investment in finding alternatives to critical metals and in developing extraction capabilities from oxides have become very necessary for a developing country like India.

India set up Indian Rare Earths Limited (IREL) in 1950 to secure strategic minerals. Now under the Department of Atomic Energy, it operates plants in Kerala, Tamil Nadu, and Odisha, and supplies materials to the nuclear, defence, space, and largely, medium and small enterprise (MSE) customers in India. It is known to have partnered with Oil India, Coal India, and NLC to pursue critical mineral assets and is exploring overseas collaborations. Internally, it aims to triple capacity and claims that by 2031–32 it can meet much of domestic demand. India produces 8 of 17 rare earths, namely lanthanum, cerium, praseodymium, neodymium, samarium, dysprosium, gadolinium, and yttrium/samarium, mainly from beach sand minerals such as monazite. IREL is mainly focused on thorium production for nuclear power plants and breeder reactors.

According to a press report, some critical minerals have been identified in Rajasthan, Chhattisgarh, and Singrauli in MP during a mineral survey by the relevant ministry, but have yet to be confirmed.

6.9 Green Hydrogen Valley.

A Green Hydrogen Valley initiative is a localised ecosystem focused on developing the entire green hydrogen value chain within a specific region. These initiatives aim to integrate green hydrogen production, storage, distribution, and end-use applications across sectors such as industry, transportation, and energy. The goal is to create a self-sustaining green hydrogen economy within the designated area, driving regional economic development and contributing to climate change mitigation.

Key aspects of Green Hydrogen Valley initiatives:

- Integrated Ecosystem: They encompass the entire hydrogen value chain, from renewable energy-powered electrolysers for production to storage and distribution infrastructure, and finally, to diverse end-use applications.
- Regional Focus: They operate within a defined geographic scope, such as a city, region, island, or industrial cluster, fostering local synergies.
- Cost Reduction: A key objective is to reduce the cost of green hydrogen production and deployment to make it competitive with other energy sources.
- Decarbonisation: They aim to decarbonise traditionally hard-to-abate sectors such as heavy transport and industrial manufacturing, including steel and cement, by providing a clean energy alternative.
- Economic Development: Green Hydrogen Valleys can create new industries, jobs, and economic opportunities in the designated region.

Many countries and regions are actively developing Green Hydrogen Valleys as part of their national strategies to achieve climate goals and transition to a hydrogen economy.

Examples of Green Hydrogen Valley initiatives:

- Andalusian Green Hydrogen Valley (Spain): This project involves building 2 new hydrogen production facilities with a combined capacity of 2 GW, aiming to establish a regional hydrogen ecosystem, according to the World Economic Forum.
- South African Hydrogen Valley: This initiative is part of the country's National Hydrogen Society Roadmap and aims to integrate various hydrogen applications into a single ecosystem according to ENGIE Impact. ENGIE is a French energy group formed by the merger of Gaz de France and SUEZ to evoke energy, modernity, and a shift towards a low-carbon, sustainable future.
- Asturian Green Hydrogen Valley: Energias de Portugal (EDP), a major global energy company leading the energy transition to renewables, is transforming a power plant into a Green Hydrogen Valley with a 150 MW electrolyser capacity, demonstrating its commitment to local hydrogen production, according to EDP Global.
- Andhra Pradesh's Green Hydrogen Valley (India): This initiative focuses on building a large-scale green hydrogen ecosystem, including electrolyser manufacturing, green hydrogen production, and infrastructure development.

As part of its push into this futuristic technology, Andhra Pradesh has set ambitious targets: to establish an electrolyser manufacturing capacity of 5 GW by 2029 and to produce 1.5 million metric tonnes of green hydrogen per annum by 2029. Other goals include creating transmission infrastructure (green energy corridors) to evacuate 25 GW of renewable energy by 2029 and reducing costs in production and deployment.

7 Blue Hydrogen

Blue hydrogen is made from fossil fuels, mostly natural gas, but the process is cleaned up by capturing CO_2, using CCUS technology. Blue hydrogen is considered clean hydrogen. The allowable carbon dioxide emission limit for blue hydrogen, from well to production, not to exceed 4 kg per kg of hydrogen produced, taking into account from well-to-wheel, as per the US regulations, which India also follows.

7.1 Carbon Capture, Utilisation and Storage (CCUS)

Capture of carbon dioxide can be either from the source of emission, such as the SMR process, cement industry, steel plants, fossil fuel-based power plants, or directly from the air. If not used on-site, the captured CO_2 is compressed and transported by pipeline, ship, rail, or truck for use in a range of applications, or injected deep into geological formations such as depleted oil and gas reservoirs or saline aquifers. Saline aquifers are underground layers of rock that are saturated with salty water, also known as brine. They are distinct from freshwater aquifers because the water they contain is not suitable for drinking or most other uses without desalination. These formations are found deep underground and are considered promising sites for carbon storage projects without contaminating freshwater bodies.

TOI, dated 2 November 2025, reported that NTPC, India, plans to drill a geological well at the Barwadih coal mine in Jharkhand to a depth of 1,200 metres to collect geological and reservoir data, with the objective of storing captured CO_2 in the CCUS programme. If the reservoir condition is found favourable, blue hydrogen can be produced from an integrated gasification combined cycle (IGCC) power plant to be set up in the coal-rich Jharkhand region, with carbon capture. Carbon capture from direct coal-fired power plant flue gas is economically not attractive because it contains a much lower CO_2 concentration (12 to 14%) at lower pressure.

CCUS can tackle emissions in hard-to-abate sectors, particularly in heavy industries such as cement, steel, and chemicals. CCUS is an enabler of least-cost low-carbon hydrogen production, which can support the decarbonisation of other parts of the energy system, such as industry and transport. CCUS can remove CO_2 from the air to balance emissions that are unavoidable or technically difficult to abate.

CCUS deployment has been behind expectations in the past, but momentum has grown substantially in recent years, with over 700 projects globally at various stages of development across the CCUS value chain. Nevertheless, even at such a level, its deployment would remain well below what is required in the Net-Zero Scenario.

CCUS can be retrofitted at the plants that may otherwise be emitting 15 billion tonnes of CO_2, around 38% of global CO_2 emissions from all sectors.

CCUS can tackle emissions in sectors that have limited options or no options to capture carbon during the production process, such as cement, steel, chemicals manufacturing, and the production of synthetic fuels for long-distance transport.

It enables the production of low-carbon hydrogen from fossil fuels, like in the steam naphtha reformation (SNR) process, a least-cost option in several regions around the world.

7.2 Technology of CO_2 Capture

CO_2 removal technologies can be broadly categorised into absorption, adsorption, and chemical reactions, each offering unique approaches to capturing and separating carbon dioxide.

Absorption involves the dissolution or chemical reaction of CO_2 with a solvent. CO_2 can be desorbed by heating the rich solution and can be captured. CO_2 can then be separately stored, and the solvent can be circulated back. Monoethanolamines (MEA) is mostly used as a solution.

Adsorption, on the other hand, relies on the adherence of CO_2 molecules to the surface of a solid material at a pressure. Adhered or adsorbed molecules on the surface can be removed by depressurising or heating. Zeolites, metal-organic frameworks (MOFs), and activated carbon are used as adsorbents.

Chemical reactions, like those in mineralisation and ammonia production, can also be used to convert CO_2 into other compounds. In a chemical reaction, a new compound is formed, which then becomes difficult to separate from.

Carbon capture from flue gas in a power plant or any combustion unit can use 2 basic pathways:

1. Precombustion,
2. Post-combustion

7.2.1 Precombustion Capture of CO_2

The goal of precombustion capture is to remove CO_2 prior to burning fossil fuels. Before burning, fossil fuels are gasified and reconstituted by air and water vapour, yielding CO_2 and H_2. Precombustion capture involves a steam reforming process that converts fuel to syngas (a mixture of carbon monoxide (CO) and H_2, followed by a WGS reaction that converts CO to CO_2 and H_2, thereby increasing total hydrogen production CO_2 produced by the process is separated using gas separation techniques, primarily adsorption. CO_2 is recovered and transported for use in industries or geological storage.H_2 is utilised as fuel for power generation. Higher partial pressure of CO_2 in the SMR process increases the recovery of CO_2 in precombustion capture. The process flow diagram is shown in Figure 23.

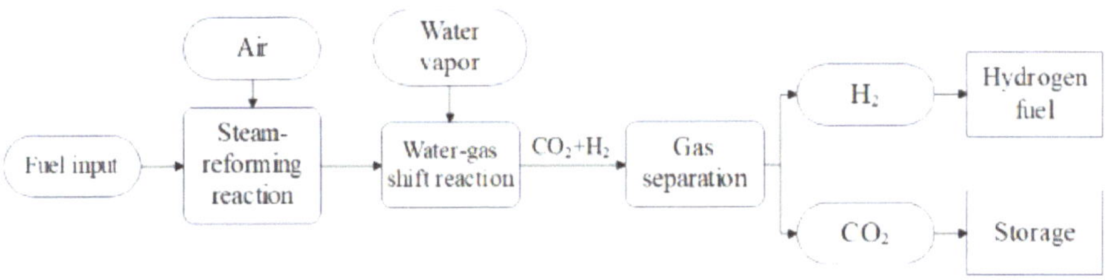

▲ Figure 23: Precombustion carbon capture

7.2.2 Post-Combustion Capture of CO_2

In post-combustion capture, CO_2 is captured from flue gas. The concentration of CO_2 in flue gas depends upon the type of fuel used, which has varying carbon content. In coal-fired power plants, carbon dioxide content in flue gas varies between 12–14%, whereas in natural gas-fired power plants, it is 3–5%. The extent of recovery depends upon the partial pressure of CO_2 in flue gas. The higher the concentration or partial pressure, the better the recovery.

In the post-combustion CO_2 removal process, flue gas undergoes dust removal, denitrification, desulphurisation, and chilling to prevent solvent breakdown. The flue gas, primarily composed of CO_2, H_2O, and N_2, is supplied counter-currently to an absorption tower from its bottom side, and solvent is fed from the top side of the tower. Flue gas and solvent come into contact with each other in an absorption tower with a packed bed, valve trays, or bubble cap trays in series. The flue gas rises up in the tower, leaving CO_2 behind, and solvent falls down absorbing CO_2 from a series of trays above. The solvent rich in CO_2, called the rich solvent, is collected at the bottom of the absorption tower and heated to desorb CO_2 in a separate tower, called the regeneration tower. Desorbed CO_2 comes out from the top of the regeneration tower, and solvent, called lean solvent, is collected from the bottom, which is cooled and circulated back to the absorption tower with the help of a circulation pump.

Nitrogen in flue gas is taken out from the top of the absorption tower. CO_2 is collected from the top of the regeneration tower, which is cleaned and dried, and then stored or transported to other destinations for use or storage.

This is a continuous process in which the solvent that absorbs CO_2 is regenerated and recirculated in a closed loop. CO_2 can either be used or undergo compression into a supercritical fluid and sent to saline aquifers or geological reserves for retention. CO_2 is compressed to a supercritical state because it allows for greater density and reduced buoyancy, making it ideal for geological storage.

Currently, solvents derived from monoethanolamine (MEA) are employed in all commercial post-combustion CO_2 capture facilities. MEA is a well-proven chemical that was created more than 70 years ago for removing CO_2 and hydrogen sulphide from natural gas streams.

Figure 24 shows a conventional CO_2 capture process used in post-combustion carbon capture.

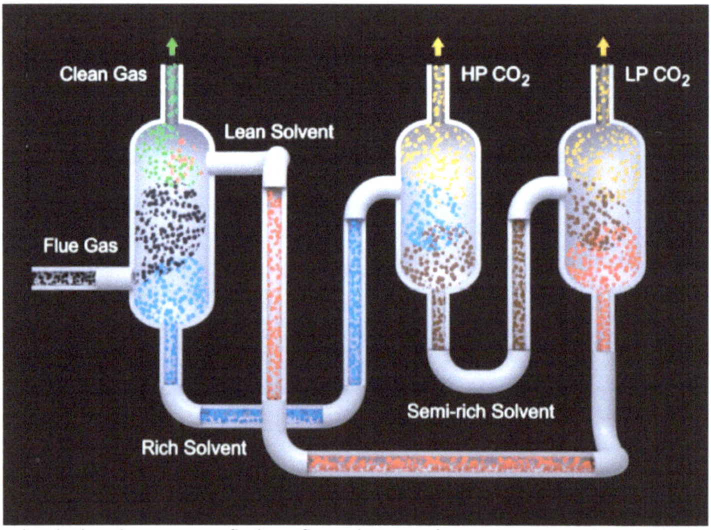

This animation depicts the two-stage flash configuration, one of several processes described in a new study detailing how EEMPA, a Pacific Northwest National Laboratory-developed solvent, can capture carbon from flue gas emitted by power plants. From left to right, EEMPA (red) first interacts with flue gas (black), where it absorbs carbon dioxide. Then, as a saturated solvent (blue), EEMPA is stripped of carbon dioxide in high and low-pressure tanks. Finally, the stripped solvent is reintroduced to the carbon dioxide absorber, where the process begins again. (Animation by Michael Perkins | Pacific Northwest National Laboratory)

▲ Figure 24: CO_2 Capture Process Post Combustion

Using CCUS alongside SMR can reduce carbon emissions from the process by about 90%. Other technologies that could be pressed into service and are already in use in some chemical industries, such as autothermal reforming (ATR). ATR has higher capture rates.

ATR is a chemical process that combines partial oxidation and steam reforming reactions to produce syngas (a mixture of hydrogen and carbon monoxide). This process utilises both the exothermic (heat-releasing) partial oxidation reaction and the endothermic (heat-absorbing) steam reforming reaction within a single reactor, achieving thermal self-sufficiency. Both processes, partial oxidation and reforming, produce synthetic gas. Their simplified reactions

When using carbon dioxide:

$$2CH_4 + O_2 + CO_2 \rightarrow 3H_2 + 3CO + H_2O$$

When using steam:

$$4CH_4 + O_2 + 2H_2O \rightarrow 10H_2 + 4CO$$

7.2.3 Oxyfuel Combustion

Oxyfuel combustion involves burning fossil fuels in pure oxygen, producing nitrogen-free flue gas. Only CO_2 and H_2O are present in oxyfuel flue gas. Flue gas condensation results in the production of a pure CO_2 stream. This way, the separation of CO_2 from flue gas becomes very easy.

Using the oxyfuel combustion method in power-generating plants can reduce flue gas and nitrogen oxide (NOx) emissions while also enhancing boiler efficiency. One major problem in such a route is the provision of pure oxygen. This adds to capital expenditure.

7.2.4 Cost of Carbon Capture

CCUS applications do not have the same cost in all places. Carbon capture costs can vary widely by CO_2 source, from USD 15-25/t CO_2 for industrial processes with concentrated CO_2 streams (such as ethanol production or natural gas processing), to USD 40-120/t CO_2 for processes with "dilute" gas streams (such as cement production and power generation). Capturing CO_2 directly from the air, where the concentration is very low, 425 ppm, is currently the most expensive approach, but could nonetheless play a unique role in carbon removal. Many CO_2 capture technologies are commercially available now, while others are still in development.

The levelised cost of CO_2 capture by sector is given in Figure 25, taken from the IEA.

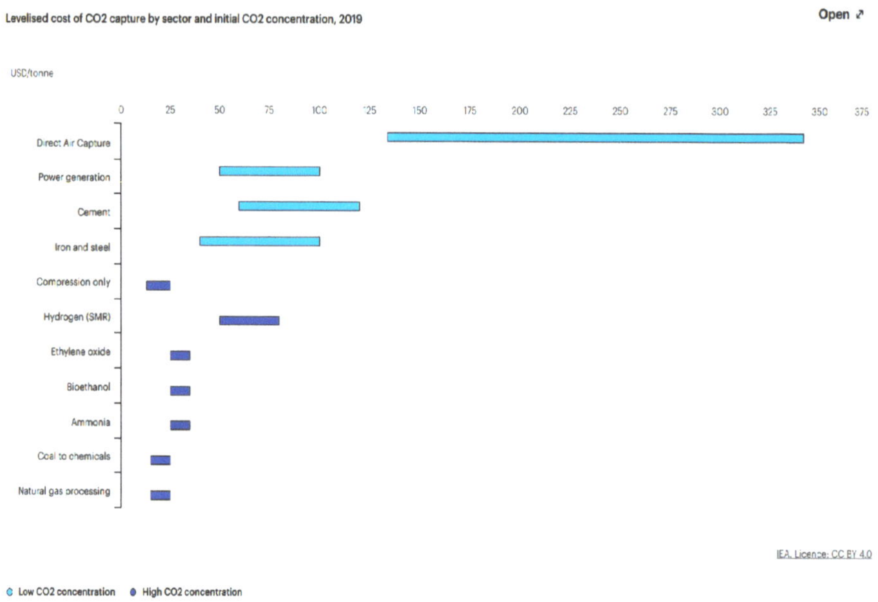

▲ Figure 25: Levelised cost of CO2 Capture by sectors

CO_2 capture costs for hydrogen refer to production via SMR of natural gas; the broad cost range reflects varying levels of CO_2 concentration: the lower end of the cost range applies to CO_2 capture from the concentrated "process" stream, while the higher end applies to CO_2 capture from the more diluted stream coming out of the SMR furnace. Cost estimates are based on the United States. All capture costs include the cost of compression.

Moving on to the cost of transport and storage, this can also vary greatly on a case-by-case basis, depending mainly on CO_2 volumes, transport distances, and storage conditions. In the United States, for example, the cost of onshore pipeline transport ranges from USD 2–14/t CO_2. In some cases, transport and storage costs can even be negative if the CO_2 is injected into oilfields to enhance production and thus generate more revenue from oil sales.

According to the Global CCS Institute's *Cost of CO_2 Storage* report, issued in December 2025, underground geological storage costs are highly site-specific. While many factors affect storage costs, the 5 main drivers are tectonic settings (onshore or offshore), reservoir quality (porosity and permeability), reservoir thickness, fracture pressure gradient, and boundary conditions.

For the onshore scenarios considered, costs were estimated to range from US \$2 to US \$ 15/tCO_2, remaining below US \$60/tCO_2 even in less favourable conditions.

In offshore scenarios, under favourable conditions, offshore costs can be close to onshore costs, ranging from US \$5-31/tCO_2. In constrained settings, it can go up to US \$147/tCO_2. Open boundaries help to reduce costs and narrow this range.

Boundary conditions are the main cost factor. The geology of a reservoir, including its tectonic setting and depositional environment, determines whether it has open or closed boundary conditions. Under open boundary conditions, where pressure can dissipate across the boundary, projects can be up to 4 times cheaper than under closed conditions, as pressure relief helps maintain injectivity and enables more CO_2 to be stored over the project's life.

Reservoir properties, while secondary to the principal cost drivers of boundary conditions (open versus closed) and setting (onshore versus offshore), are significant indicators of likely cost. Site selection for good quality, thick reservoirs with a high fracture pressure gradient (FPG) tolerance results in the most cost-effective storage.

The geological storage cost estimates presented in Figure 26 are based on a limited set of site-specific scenarios. They are intended to illustrate how these drivers influence the cost per tonne of CO_2 stored. This is taken from the Global CCS Institute report, December 2025.

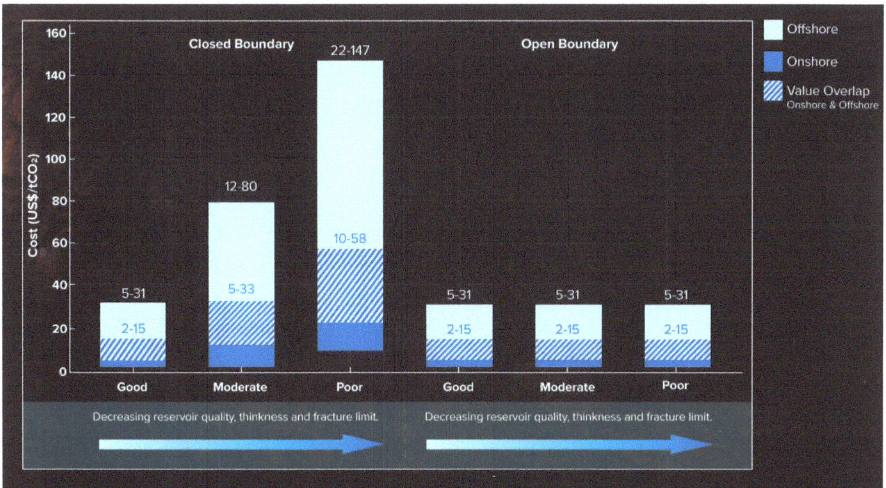

This figure compares onshore (dark blue) and offshore (light blue) storage under closed and open boundary conditions. Costs escalate sharply in closed systems as formation quality, thickness, and fracture pressure gradient decrease — reaching up to US\$147/tCO_2 offshore in poor-quality reservoirs. In contrast, open systems remain stable, with both onshore and offshore clustered between US\$2-31/tCO_2. Boundary conditions, more than reservoir quality alone, are the dominant driver of storage cost variability.

▲ Figure 26: CO_2 Geological Storage cost estimate

7.2.5 Challenges in CCUS Technology

Carbon capture technologies have become integral to global efforts aimed at reducing carbon dioxide emissions from industrial sources. However, these technologies face significant challenges and limitations, casting doubt on their feasibility and effectiveness.

One of the primary challenges associated with carbon capture systems is their high cost. The financial burden of implementing carbon capture technologies is substantial. A study highlights that the cost of capturing CO_2 from power plants is often considered economically unviable, with estimates ranging from $50 to $100 per tonne of CO_2. These costs are driven by the need for advanced materials, significant energy inputs, and the development of extensive infrastructure. Additionally, transportation and storage expenses contribute to the overall cost. The considerable initial and operational expenses associated with these technologies raise concerns about their widespread adoption, particularly in developing countries where financial resources are often limited.

Another major obstacle is the long-term storage of collected carbon dioxide. One of the most frequently utilised methods to store CO_2 is through geological storage, which entails pumping captured CO_2 underground into rock formations.

Although this technique has shown potential, questions remain regarding the stability and security of CO_2 storage. CO_2 leakage from underground storage facilities has detrimental effects on the environment. It is reported that when CO_2 leaks into potable groundwater, it can increase the water's acidity, leading to the dissolution of minerals. This process may release hazardous elements, such as lead and arsenic, into the water, contaminating water sources and posing a serious threat to aquatic life.

The Global CCS Institutes track 26 operating facilities around the world that can capture 40 million tonnes of CO_2 per year, and 37 under construction or development (reference "*The Hydrogen Revolution*" by Marco Alvera) as of 2021. Many of the projects are in the US, where CO_2 is sold to oil companies for enhanced oil recovery (EOR) in oil fields. Geological storage sites include the Humber region (Zero Carbon Humber) in the US and one of the east coasts of northern England (Net-Zero Teesside), both of which have the North Sea as the final destination.

CCUS is the most cost-effective way to curb emissions in steel and chemical manufacturing. It is the only mature technology to curb emissions from cement production. Climate scientists say that we will need this technology to reduce our net emissions, and we may well need CCUS to suck CO_2 out of the air to achieve our target of restricting the rise of global temperature below 2ºC by the end of the century, because we are left with limited space in the atmosphere to accommodate CO_2, which is about 800 Gt of carbon budget only.

Since 2017, there have been dozens of announcements for CCUS facilities, mostly in Europe and the US. But even if these projects go ahead, global CO_2 capture capacity will still account for only about 0.3% of global emissions.

Today, 70% of the CO_2 molecules accumulated in the atmosphere since the pre-industrial period (1860–90) originated from developed nations. This happened due to a lack of diligence by developed nations to protect the environment, despite the fact that a science journal, back in 1912, had cautioned that excess CO_2 emissions can be a threat to human safety. Today, there is no room left in the atmosphere for developing nations to safely accommodate CO_2. Though China is currently emitting the most over the past 5 years, but at the same time, it shows due diligence in protecting the environment.

It has been estimated that, over the coming years, depending on where you are, blue hydrogen is likely to be the cheapest form of clean hydrogen, costing only about $2.5 per kilogram with the application of CCUS technology.

8 Global Hydrogen Production & Demand

8.1 Global Hydrogen Production Scenario

Global hydrogen demand reached 97 million tonnes (Mt) in 2023. India is among the top consumers of hydrogen, ranking fourth globally at 6 Mtpa. While demand for hydrogen is growing, much of it is met through traditional methods that use fossil fuels. Meeting climate goals requires a shift from fossil-fuel-based hydrogen production to low-emission methods.

There is an increasing focus on using hydrogen in new applications, particularly in heavy industries and long-distance transport. Many governments are setting targets for hydrogen production for use in various industrial sectors.

Almost all of the world's current H_2 production, more than 95 Mt/yr, is produced from fossils and used for non-energy purposes. These uses mainly involve removing sulphur from refined products, upgrading products in refineries, producing ammonia and methanol, and the direct reduction of iron (DRI). DRI is called sponge iron, which is the feedstock for steel manufacturing, similar to pig iron from a blast furnace feeding to steel plants. The DRI process uses hydrogen to reduce iron ore instead of coal, thereby reducing CO_2 emissions.

Global hydrogen production is expected to increase significantly by 2050, with forecasts ranging from 500 to 660 million tonnes per year. This growth is driven by the need to decarbonise various sectors, particularly hard-to-abate industries like steel, cement, chemicals, and heavy-duty transportation.

Here's a more detailed look at the forecasts made by various agencies for the hydrogen demand that is likely to be reached by the middle of the century.

- IEA**:** Projects a need for 520 million tonnes of clean hydrogen annually by 2050 to achieve net-zero emissions, according to the IEA.
- Bloomberg NEF**:** Estimates a similar figure of 500 million tonnes by 2050.
- IRENA**:** Forecasts 614 million tonnes of clean hydrogen by the middle of the century, in their 1.5°C scenario.
- Hydrogen Council**:** Expects 660 million tonnes to be produced by mid-century.

- **Deloitte:** Highlights that decarbonising hard-to-abate sectors could require a six-fold increase in global hydrogen use, reaching nearly 600 million tonnes by 2050.
- Hydrogen Insight: Notes that if the global energy system shifts more towards net-zero targets, hydrogen could play a larger role in the energy mix.

While the figure of forecasted hydrogen production varies depending on the source and scenario, the overall trend is clear: a substantial increase in global hydrogen production is expected by 2050, driven by the need for decarbonisation and hydrogen's capability to play a key role in a net-zero future.

The world's total future hydrogen demand is broadly divided into 3 categories:

1. Decarbonising existing non-energy sectors — replacing unabated fossil fuels with lower-emission alternatives in non-energy sectors like steel and cement.
2. Fuel directly switching to hydrogen in the energy sector—retrofitting and modification of infrastructure shall be needed.
3. Hydrogen to its derivatives, like green ammonia and green methanol, where new infrastructure has to be established.

Energy Transition Outlook 22, by DNV, estimates that non-energy uses of hydrogen are expected to continue to grow slowly until the mid-2030s, declining thereafter to current levels by the mid-century, with falling demand mainly associated with the oil products, and the associated use of hydrogen in refineries. By 2040, hydrogen demand for energy will catch up with non-energy use of hydrogen. In 2050, only 30% of the global hydrogen supply will be used for non-energy purposes. 39% will be direct use of hydrogen for energy, while 31% will be used to produce green ammonia, green methanol, or e-fuel for energy end users.

Figure 27 shows the predicted hydrogen demand by 2050 across different industrial sectors, based on *Hydrogen Energy Forecast 2050* by DNV in the Energy Transition Outlook 2022.

On a regional level, China accounted for almost 30% of global hydrogen production, reflecting large domestic demand for refineries and the chemical industry. In 2022, more than 70% of global production was in China, the United States, the Middle East, India, and Russia, in descending order, by share of production.

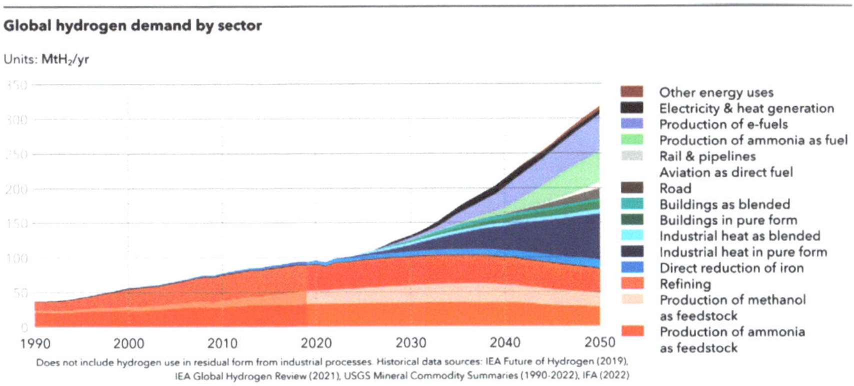

▲ Figure 27: Global Hydrogen Demand by year

8.2 Global Hydrogen Demand

McKinsey & Company, in its report Global Energy Prospective 23, Hydrogen Outlook, has predicted the demand for clean hydrogen by 2050 across 5 scenarios. These scenarios have been defined in Figure 28.

Looking at the global awareness and commitment, as per the author's view, the current trajectory scenario can be considered more realistic, and the Further Acceleration scenario is much expected.

Nearly all hydrogen consumed today is grey hydrogen, approximately 90+ million tonnes per annum, and is driven mainly by refineries and fertiliser industries. New applications of hydrogen—for instance, in steel, in the production of synthetic fuels, and in heavy road transport—may emerge from 2030 onwards. Demand for grey hydrogen is projected to decline as demand for clean hydrogen rises, as green hydrogen costs eventually become more competitive. By 2050, clean hydrogen demand could account for 73–100% of total hydrogen demand, with the balance met by grey hydrogen, depending on the scenario.

After 2040, private and public-sector commitments are projected to drive the uptake of clean hydrogen and hydrogen-based fuels in emerging applications in the Further Acceleration and Achieved Commitments scenarios. Potential mechanisms that would be required to support demand growth of clean hydrogen and hydrogen derivatives in these applications include the implementation of pricing mechanisms, quotas on sustainable fuels in aviation, or CO_2-reduction targets in maritime transportation.

Some geographies, such as the European Union and the United Kingdom, are expected to fully phase out grey hydrogen by 2050 in all scenarios except Fading Momentum. Figure 28 below shows the trend of rising clean hydrogen demand and plummeting grey hydrogen across different scenarios, as estimated by McKinsey Energy Solutions in the Global Hydrogen Review 2023. Scenarios are defined in Figure 28.

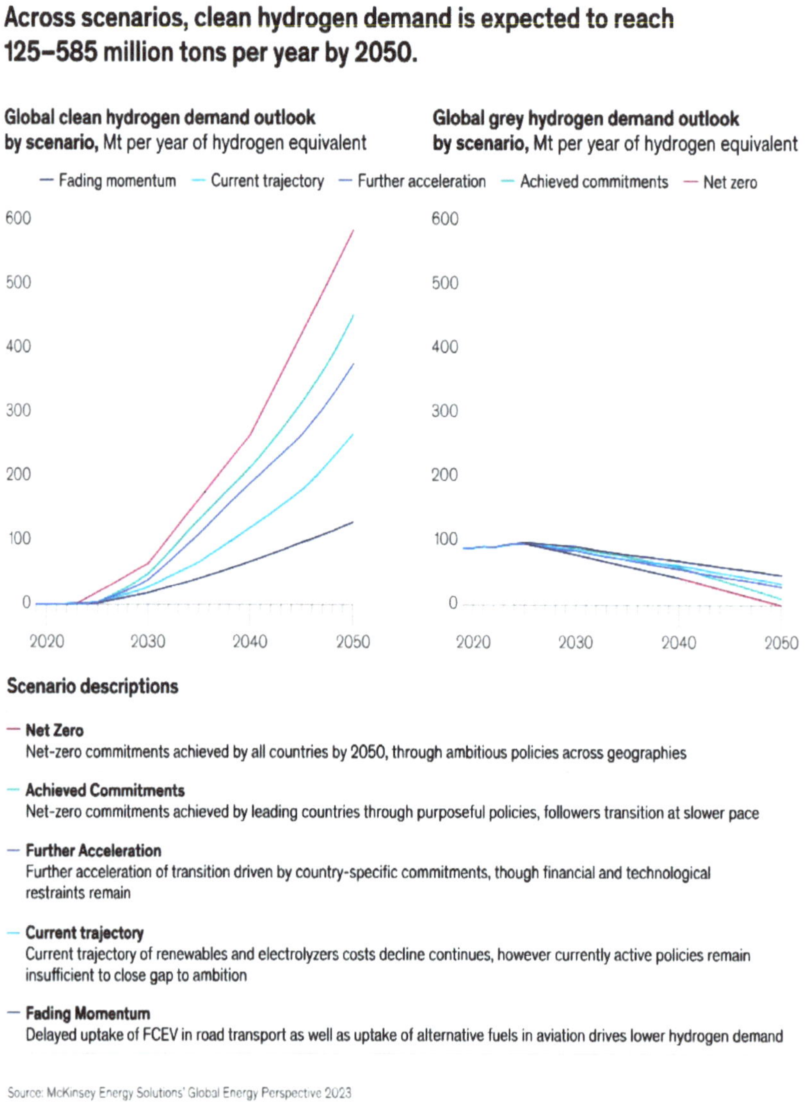

▲ Figure 28: Clean Hydrogen Demand Outlook.

8.3 Hydrogen Demand in the Mobility Sector

By 2040, clean hydrogen could play a larger role in new applications—especially in mobility, which is expected to be the largest "newcomer" for clean hydrogen demand in McKinsey's *Further Acceleration Scenario*. In the current trajectory, hydrogen demand in the mobility sector is expected to reach 106 million tonnes per annum in 2050. Applications could range from fuel cell electric vehicles in long-haul, fuelling heavy-duty trucking to synthetic kerosene in aviation. The second-largest newcomer is expected to be hydrogen used for (mainly industrial) heating, displacing natural gas.

Figure 29 shows the expected growth in hydrogen demand by year for different sectors in different scenarios defined earlier in Figure 28. According to the vertical bar chart, clean hydrogen demand (both green and blue together) is estimated to account for about 98% of total hydrogen demand (463–530 million tonnes) by 2050. Grey hydrogen shall be very low, only 2%.

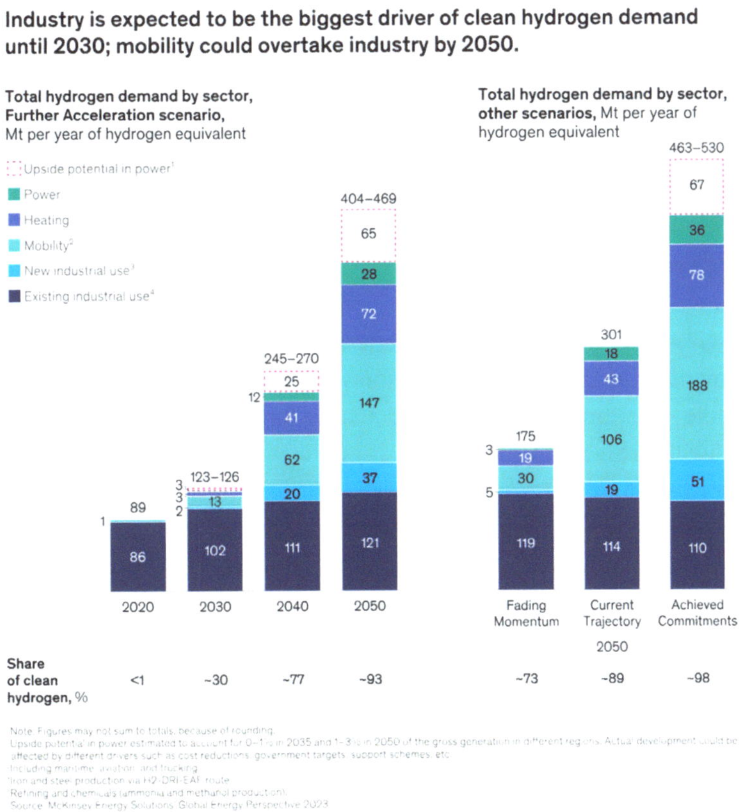

▲ Figure 29: Expected Growth in Hydrogen Demand

According to McKinsey & Company's *Global Energy Perspective 2023: Hydrogen Outlook*, by 2050, green hydrogen is expected to dominate the global hydrogen supply mix, with a share of 50-65% across scenarios. This is because the cost reductions in renewable energy and electrolysis make the production route more cost-competitive. Blue hydrogen is projected to account for the next-largest share of supply, between 20 and 35%.

The ratio of blue to green hydrogen production is expected to differ significantly by region, driven mainly by cost factor developments. Blue hydrogen production is projected to be concentrated in regions with cost-competitive natural gas and CCUS, such as the Middle East and North America. By 2050, blue hydrogen production could require as much as 500 billion cubic metres of natural gas (between 10 and 15 percent of global natural gas demand in the Further Acceleration scenario) and the capacity to capture and store 750 to 1,000 megatons of CO_2.

Green hydrogen production is projected to have a larger share in regions with abundant, cost-competitive renewable resources. The production of green hydrogen could potentially be constrained by a lack of renewable power. Globally, approximately a quarter of renewable electricity generation (around 14,000 terawatt-hours) could be required to produce the green hydrogen needed by 2050 in the Further Acceleration scenario. Further increases in green hydrogen uptake can be achieved through large-scale investments and the deployment of at-scale manufacturing of electrolysers with cost competitiveness.

Figure 30 shows the green, blue, and grey hydrogen uptake by different regions. Green hydrogen supply uptake is expected to be similar across all regions, while blue hydrogen production will be specific to certain geography. The Middle East and North America, which have an abundance of gas, will favour blue hydrogen production.

According to the *Hydrogen Insight* report by McKinsey & Company, issued in February 2021, there are 228 clean hydrogen projects globally across the value chain. Of these, 17 are already announced giga-scale production projects, with the largest in Europe, Australia, the Middle East, and Chile. More than 1 GW of renewable energy and over 2 lakh tonnes of blue hydrogen per year are included in gigawatt-scale projects.

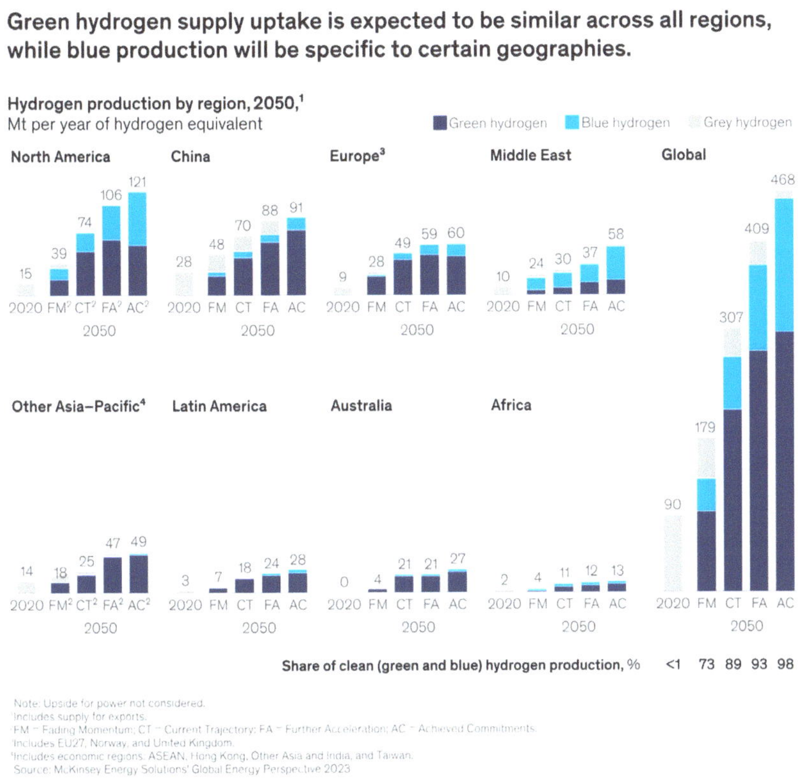

▲ **Figure 30:** Green and Blue Hydrogen Supply

Europe leads globally in the number of announced hydrogen projects, with Australia, Japan, Korea, China, and the USA following as additional hubs. Of all announced projects, 55% are located in Europe. The announced projects cover the entire hydrogen value chain, including source, midstream, and downstream. In major demand centres such as Korea, Japan, and Europe, the focus is on industrial use and transport application projects. Japan and Korea are strong in road transport applications, green ammonia, LH_2, and liquid organic hydrogen carrier (LOHC) projects. Europe champions multiple integrated hydrogen economy projects. A tally of project announcements, investments required to meet government production targets, and spending projections across the value chain totals more than USD 300 billion through 2030.

Figure 31 below shows global hydrogen projects across the value chains, taken from McKinsey & Company.

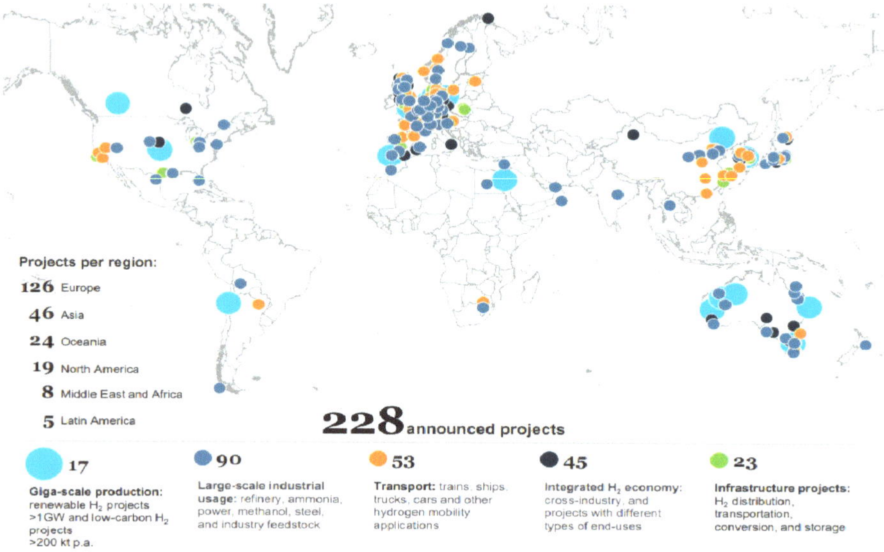

▲ Figure 31: Global hydrogen projects across value chain

Clean hydrogen projects that reached final investment decision (FID) have seen a dramatic seven-fold increase in value, from USD 10 billion in 2020 to USD 75 billion in 2024. Key factors driving successful projects to FID include clear and effective incentives, demand-side visibility that drives offtake, and strong industrial policy driving costs down. Over the years, a larger portion of projects has shifted from announcements to more advanced stages. Between 2020 and 2024, investments made in front-end engineering design (FEED) stage projects increased by a factor of 20.

Natural attrition drives industry maturation by phasing out less viable projects and prioritising those with the highest potential. Similar trends have been observed in the early years of the wind and solar industries prior to reaching maturity.

A key sector-specific challenge for the hydrogen industry is uncertainty, associated with a number of regulatory frameworks that impede project bankability. McKinsey & Company, Feb 2021, has estimated that, to accelerate the global energy system's decarbonisation, an 8-fold increase in hydrogen investment is required by 2030, compared with the current investment of USD 75 billion past FID.

9 Storage and Transportation of Hydrogen

Hydrogen storage is a key enabling technology for the advancement of hydrogen and fuel cell technologies in applications, including stationary power, portable power, and transportation. Hydrogen has the highest energy per unit mass of any fuel; however, it has a low energy per unit volume. It requires the development of advanced storage methods with the potential for higher energy density.

Hydrogen storage and transportation involve various methods depending on the scale and application. Physical storage options include compressed gas in high-pressure tanks and liquid hydrogen in cryogenic tanks. Chemical storage involves binding hydrogen to materials such as ammonia or methanol to facilitate easier handling. Transportation methods include pipelines, trucks (both compressed gas tube trailers and liquid tankers), and potentially, ships or barges.

9.1 Physical Storage

Compressed Hydrogen:

Hydrogen gas is compressed to high pressures (350-700 bar) and stored in durable tanks made of steel, aluminium, or composite materials. This method is suitable for various applications, including fuel cell vehicles.

Liquid Hydrogen - Hydrogen is cooled to extremely low temperatures (-252.8°C) to become a liquid, significantly reducing its volume. The gas-to-liquid ratio of hydrogen is approximately 1:848, meaning one volume of liquid hydrogen expands to 848 volumes of gas at atmospheric temperature and pressure. This significant expansion is due to hydrogen's low boiling point of −253°C at 1 atmosphere, and it is a critical factor in designing safe storage and handling systems, such as those with pressure relief valves. Cryogenic tanks are used to store and transport liquid hydrogen, often for large-scale applications.

Solid Hydrogen Storage - Hydrogen can be stored by adsorbing or absorbing it in solid materials like metal hydrides.

An alternative to compressed and liquefied hydrogen is materials-based storage. Here, solids and liquids that are chemically able to absorb or react with hydrogen are used to bind it. This includes creating metal hydrides from elements such as palladium, which can absorb hydrogen 900 times its own volume, as well as magnesium, aluminium, and certain alloys.

Using ammonia, a compound of hydrogen and nitrogen, as a carrier for hydrogen is arguably the optimal potential. Liquid ammonia has a higher volumetric energy density than liquid hydrogen. Liquid ammonia stores approximately 12.7 MJ/L, while liquid hydrogen stores about 8.5MJ/L. Liquid ammonia is denser and can be stored at a lower temperature (-33°C) than liquid hydrogen (-253°C). This makes ammonia easier to transport and store.

Hydrogen is converted to ammonia, transported to its destination, and then "cracked" to release the hydrogen at its point of use. Synthesis and cracking of ammonia are costly and energy-intensive processes. It consumes around 4.3 kWh/kg of H_2 for synthesis to ammonia and 10 kWh/kg of H_2 for cracking, apart from transportation and other losses. More than one-third of hydrogen energy is lost in synthesis and cracking only. Therefore, using ammonia directly as fuel, without cracking, is the preferred option.

It has been suggested that ammonia could become the "workhorse" of a future hydrogen society, especially where direct use of hydrogen is not an option. Countries such as Japan and South Korea are looking to import ammonia. For example, MHI's shipbuilding division is designing a vessel that will be powered by ammonia and also ship ammonia.

LOHCs, such as methanol, are also used as hydrogen storage and carriers, similar to ammonia.

Figure 32 shows various methods of hydrogen storage. While physical-based storage is a well-developed technology, material-based storage is still in the developing stage.

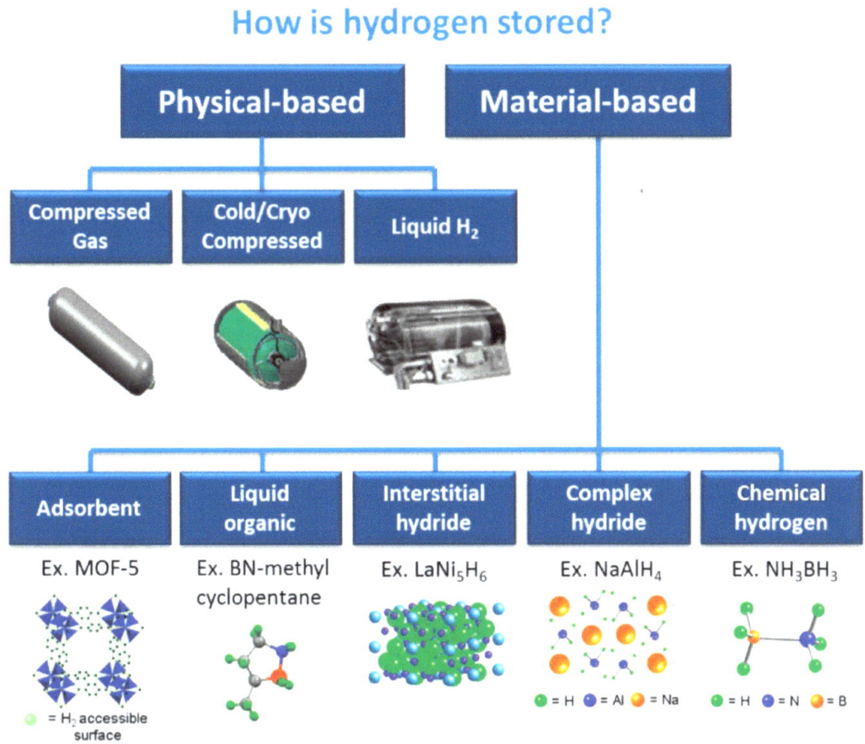

Hydrogen can be stored physically as either a gas or a liquid. Storage of hydrogen as a gas typically requires high-pressure tanks (350–700 bar [5,000–10,000 psi] tank pressure). Storage of hydrogen as a liquid requires cryogenic temperatures because the boiling point of hydrogen at one atmosphere pressure is −252.8°C. Hydrogen can also be stored on the surfaces of solids (by adsorption) or within solids (by absorption).

▲ Figure 32: Hydrogen Storage by different pathways

9.2 Hydrogen Carriers

Carriers are a unique way to deliver hydrogen by hydriding a chemical compound at the site of production and then dehydriding it either at the point of delivery or once it is on board the fuel cell vehicle. This method of hydrogen delivery is still in the early stages of research and development, and as of yet, it has not been shown to be energy or cost-efficient. It is beneficial that these hydrides can be used as an energy source in their form.

Types of Hydrogen Carriers

- Ammonia (NH3): Ammonia can be liquefied under moderate pressure and temperature, making it relatively easy to handle. It is also a potential fuel source for various applications. Its energy density is 12.7 MJ/L.
- Metal Hydrides: These compounds store hydrogen by absorbing it into a metal lattice.
- LOHCs: These are organic compounds that can absorb and release hydrogen through chemical reactions. Examples include dibenzyltoluene, methylcyclohexane, and methanol. LOHCs offer higher volumetric energy density than compressed or liquefied hydrogen and can be transported using existing liquid fuel infrastructure. Methanol volumetric energy density is 15.6-16 MJ/L, against liquid hydrogen 8.5-9.1MJ/L.

LOHCs can store more hydrogen per unit volume compared to other methods. It can potentially utilise existing liquid fuel infrastructure for transportation and storage, and therefore, it can reduce the cost associated with transportation, storage, and distribution. LOHCs can be handled more safely than pure hydrogen gas.

Examples of LOHC Use:

- Industrial Heat: LOHCs can be used to transport hydrogen to industrial sites that require high-grade and high-intensity heat, such as steel or cement production.
- Heavy Mobility: LOHCs can enable the use of hydrogen in heavy-duty vehicles like trucks and buses, which require long ranges and rapid refuelling.

Pipeline transportation of liquid hydrogen (LH_2), ammonia (NH_3), and LOHC is competitive for long-distance hydrogen transportation. Gaseous hydrogen is not suitable for long-distance shipping mainly because of its low energy density and high compression cost. Suppliers can liquefy hydrogen or convert it to ammonia, or bind it to a liquid organic hydrogen carrier. If every step of the value chain uses green energy and hydrogen is produced from low-carbon sources, all 3 carriers are considered green.

The long-term optimal choice of carriers depends on a range of factors. LH_2 is most efficient when the destination requires liquid or high-purity hydrogen, and offers benefits when hydrogen needs to be distributed by truck after landing at port. This is typically the case for hydrogen refuelling stations for cars or trucks. NH_3 and LOHC do not require dehydrogenation or cracking to convert into gaseous hydrogen, if these are used as fuel, which not only saves costs but also avoids purity challenges caused by carrier residues.

LH_2's main drawback is its relatively low volumetric energy density compared with ammonia and methanol, which limits the amount of hydrogen transportation per ship due to its larger liquid volume, and the boil-off losses that occur with every day of storage. While liquefaction is a proven and commercialised technology, shipping and large-scale storage of liquid hydrogen, which requires suppliers to manage the boil-off losses, remain in the early stages of deployment.

Ammonia is the straightforward answer for end-uses that need ammonia as a feedstock and can therefore avoid the need to crack NH_3 back into hydrogen, such as for fertiliser, shipping fuel, co-firing, or ammonia combustion for power generation. Suppliers are also considering this approach for other hydrogen use cases. Ammonia benefits from a higher volumetric energy density than does liquid hydrogen, and thus, suppliers can ship it more cost-effectively than LH_2 using commercially available ammonia ships.

However, the drawbacks of using ammonia as a hydrogen carrier are the high cost of cracking it back to hydrogen and the achievable purity of hydrogen from cracking. Furthermore, because ammonia is toxic, it may face handling and storage restrictions in residential areas, as well as limited options for inland distribution.

Liquid organic hydrogen carriers can use existing diesel infrastructure and safely store hydrogen over long periods without loss. The main drawbacks of LOHC are the novelty of the dehydrogenation process, which requires large amounts of heat to release hydrogen from the carrier.

9.3 Hydrogen Loss in Transportation

Energy loss in converting hydrogen to its carrier and reconverting it back for use in the form of hydrogen is given in Figure 33. The loss in transportation and distribution has not been taken into account, which will add to the total loss.

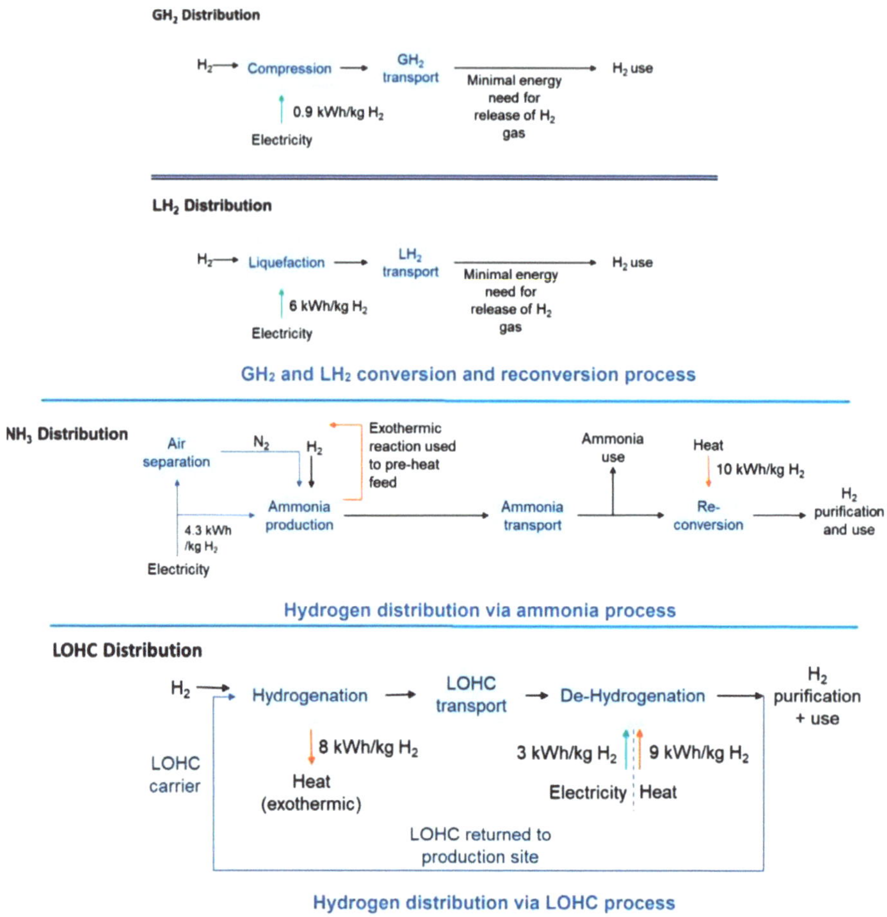

Silvian Baltac, Matt Wilson, Conor O'Sullivan, Cor Leguijt, Mart Beeftink, Isabel Nieuwenhuijse, Antonia Mattos, Diederik Jaspers, 'IEAGHG, "Blue Hydrogen: Beyond the Plant Gate", 2022-06, August 2022. This study was developed in parallel with 'Low-Carbon Hydrogen from Natural Gas: Global Roadmap'. https://ieaghg.org/publications/blue-hydrogen-beyond-the-plant-gate/
JA LI Post 04/12/2024

▲ Figure 33: Energy loss in Hydrogen Conversion and Reconversion.

9.4 Pipeline Transport of Hydrogen

Hydrogen can be transported via pipelines. Pipelines are a cost-effective method for transporting large volumes of hydrogen over long distances, especially where demand is high and relatively stable. However, technical challenges in handling exist. Hydrogen embrittlement, high diffusivity, and safety concerns related to ignition add cost to their mitigation. Natural gas pipelines are not suitable for transporting hydrogen in pure form; however, 20% blending with natural gas is practised by many industries, keeping the risk as low as practicable.

The high initial capital cost of hydrogen pipeline construction for transportation constitutes a major barrier to expanding hydrogen pipeline delivery infrastructure. Transporting gaseous hydrogen, blended with natural gas via existing natural gas pipelines, is a low-cost option for delivering large volumes of hydrogen, with limitations on pipeline pressure, temperature, and concentration.

Research today focuses on overcoming technical challenges related to pipeline transmission, including:

- Hydrogen embrittlement in steel and the welds, which are used to fabricate the pipelines.
- The need to control hydrogen permeation and leaks.
- The need for lower-cost, more reliable, and more durable hydrogen compression technology.

One possibility for rapidly expanding the hydrogen delivery infrastructure is to adapt part of the natural gas delivery infrastructure to accommodate hydrogen. Converting natural gas pipelines to carry a blend of natural gas and hydrogen (up to about 15% hydrogen as per the US Department of Energy) may require only modest modifications to the pipeline. Converting existing natural gas pipelines to deliver pure hydrogen may require more modifications or replacement.

GTI, through risk analysis, has concluded that up to 20% hydrogen can be added to natural gas in pipelines without a significant increase in societal risk, provided modest risk management is implemented. According to NaturalHy (a project focused on defining the conditions under which hydrogen can be added to natural gas in the existing natural gas system), the existing pipeline material can handle up to a 50% hydrogen blend with enhanced risk management. As per the societal risk study, blending H_2 into natural gas beyond 20% is not recommended. Societal risk governs so far as implementation is concerned.

Potential solutions include using fibre-reinforced polymer (FRP) pipelines for hydrogen distribution only, because FRP is more likely to permeate with the rise of temperature and pressure. The installation costs for FRP pipelines are about 20% lower than those for steel pipelines because FRP can be supplied in sections much longer than steel, minimising welding requirements.

Hydrogen pipeline transportation is cheaper than electricity transmission lines. Hydrogen can transport 10 times the energy at one-eighth the cost associated with electricity transmission lines. Furthermore, hydrogen pipelines have a longer lifespan than electricity transmission lines and a lower maintenance cost. Pipelines offer dual functionality, serving as both a transmission and storage medium for green energy.

While distribution networks cover regional and last-mile transport, onshore and subsea transmission pipelines can move hydrogen over distances of 500 to 5,000 kilometres or more. Pipelines can achieve extremely low-cost H_2 transport, compared with alternative transportation modes, especially where retrofits of existing infrastructure are possible. For example, retrofitting pipelines can save 60–90% of the cost of greenfield pipeline development.

Typical capex costs for onshore hydrogen transmission networks, including compression, will range between USD 0.6 to 1.2 million per km for retrofit, and USD 2.2 to 4.5 million per km for newly built H_2 pipelines, resulting in H_2 transport costs rising to USD 0.13–0.23/kg/1000km, as reported by McKinsey & Company.

Offshore/subsea hydrogen transmission pipelines cost 1.3 to 2.3 times higher due to the specific challenges and conditions of subsea pipeline construction and operation for both new projects and retrofits. Distribution pipelines are substantially cheaper than transmission pipelines (roughly 15% of transmission pipeline costs), given their smaller diameter and lower pressure.

9.5 Compression of Hydrogen.

Compression of hydrogen is energy sensitive. It takes about 8 times more work energy to compress (isothermal compression) hydrogen than that of methane per unit of mass.

The essential reason is that a kilogram of hydrogen contains 8 times as many molecules as a kilogram of methane (because the mass of a hydrogen molecule is about 1/8 of the mass of a methane molecule).

If we assume, for the sake of argument, that the compression is isothermal (constant temperature, T), the work needed (W) to compress a sample of N molecules of an ideal gas from pressure P_1 to pressure P_2 is

$W=N.Bk.T.\ln(P_2/P_1)$ (Bk=Boltzmann's constant)

So if hydrogen and methane are assumed to be ideal, 8 times more work would be required per kilogram for hydrogen than for methane, but at such high pressures, the gases are far from ideal. Intermolecular forces and the finite volumes occupied by molecules are significant and different for different gases. That would account for why the ratio of work needed is not exactly 8:1.

The energy required for the compression of hydrogen in cases of adiabatic, multistage, and isothermal compression is given in Figure 34. This is taken from a study report by Ulf Bossel and Baldur Eliasson, Corporate Research, ABB Switzerland.

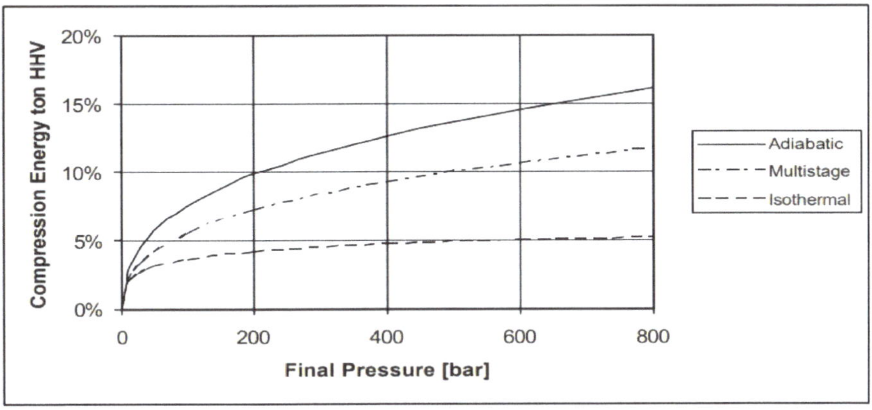

▲ Figure 34: Energy required for the compression of hydrogen compared to its HHV

9.5.1 Materials of Construction of the Hydrogen Pipeline

The following considerations are taken into account when selecting materials for the hydrogen pipeline and laying it.

1. A critical factor in selecting materials for hydrogen pipelines is their susceptibility to hydrogen embrittlement, which can lead to premature failure.
2. Materials should have low hydrogen permeability to minimise leakage and ensure safety.
3. Pipelines must be resistant to corrosion from the environment and the hydrogen itself.
4. The cost of materials and installation is a significant factor in pipeline design and construction.
5. Standards and regulations must be followed. As per Canadian Standard Regulation (CSA), the same standard and regulation may be followed as for natural gas, so long as the laying down of a hydrogen pipeline is involved.

For hydrogen pipelines, common materials include stainless steel, certain alloy steels, and fibre-reinforced polymers (FRP). Austenitic stainless steels, like 316/316L, are favoured for hydrogen service due to their compatibility and resistance to hydrogen embrittlement, a phenomenon in which hydrogen weakens the metal. This class of stainless steel is suitable for both liquid and gaseous hydrogen applications.

Certain low-alloy ferritic steels are also used, but their suitability depends on factors like temperature and pressure. Hydrocarbon pipelines use this type of steel; therefore, carrying hydrogen or hydrogen-blended gas above a certain mix, pressure, and temperature in this type of pipeline is not recommended.

According to case studies on thermal engineering reported in *ScienceDirect*, the upper limit for H_2 blending in natural gas is not fixed and depends on pipeline materials, equipment, and local regulations. High-pressure grid tests show viability up to 80 barg (1,160 psig) with 20% H_2 blending with societal risk as low as reasonably practicable. However, higher pressure and H_2 content accelerate fatigue/embrittlement in some steels (like x80), requiring pressure derating. Burner components (like boilers) need careful monitoring, as high heat loads with 20–40% H_2 can cause overheating above 380°C.

FRP pipelines offer a lighter and potentially more cost-effective solution. They are constructed with multiple layers, including a barrier layer, protective layers, and composite layers of glass or carbon fibres, providing high strength and flexibility. But it becomes permeable above a certain temperature, leading to leakage.

Today's hydrogen pipelines are often designed under the American Society of Mechanical Engineers (ASME) B31.12 standard, originally developed for short-distance, small-diameter onshore pipelines. While these standards address hydrogen embrittlement, they are widely regarded as overly conservative for future high-pressure transmission, particularly for repurposed natural gas pipelines. In response, there are plans to integrate hydrogen-specific requirements into the broader ASME B31.8 gas transmission and distribution standard update for 2026, supported by a Consensus Engineering Requirements project, completed in 2024. This update aims to streamline hydrogen pipeline design, facilitate the repurposing of existing infrastructure, and harmonise regulations relating to different gases.

The global hydrogen pipeline network is rapidly expanding, with a significant increase in both the number of projects and the volume of hydrogen production capacity that has reached FID. Notable progress on projects towards more advanced phases, with investments past FID, is growing substantially. This expansion is driven by growing government support, increasing demand from various sectors, and the development of national and regional hydrogen backbone networks.

Hydrogen pipeline projects, including new pipelines and repurposed natural gas pipelines, totalling around 37,000 km, are planned globally by 2035. However, less than 6% of this (by length) has reached FID or is under construction, which is equivalent to 2.7% of the natural gas pipelines currently under construction. If all announced hydrogen pipeline projects proceed, they would represent about 3.4% of the current global gas transmission network.

Subsequent to the energy crisis in Europe arising due to the Russia-Ukraine war, most European nations are going for alternative sources of natural gas, mostly hydrogen. A piping network of Europe, published recently in January 2025, by *Global Energy Monitor*, is placed in Figure 35.

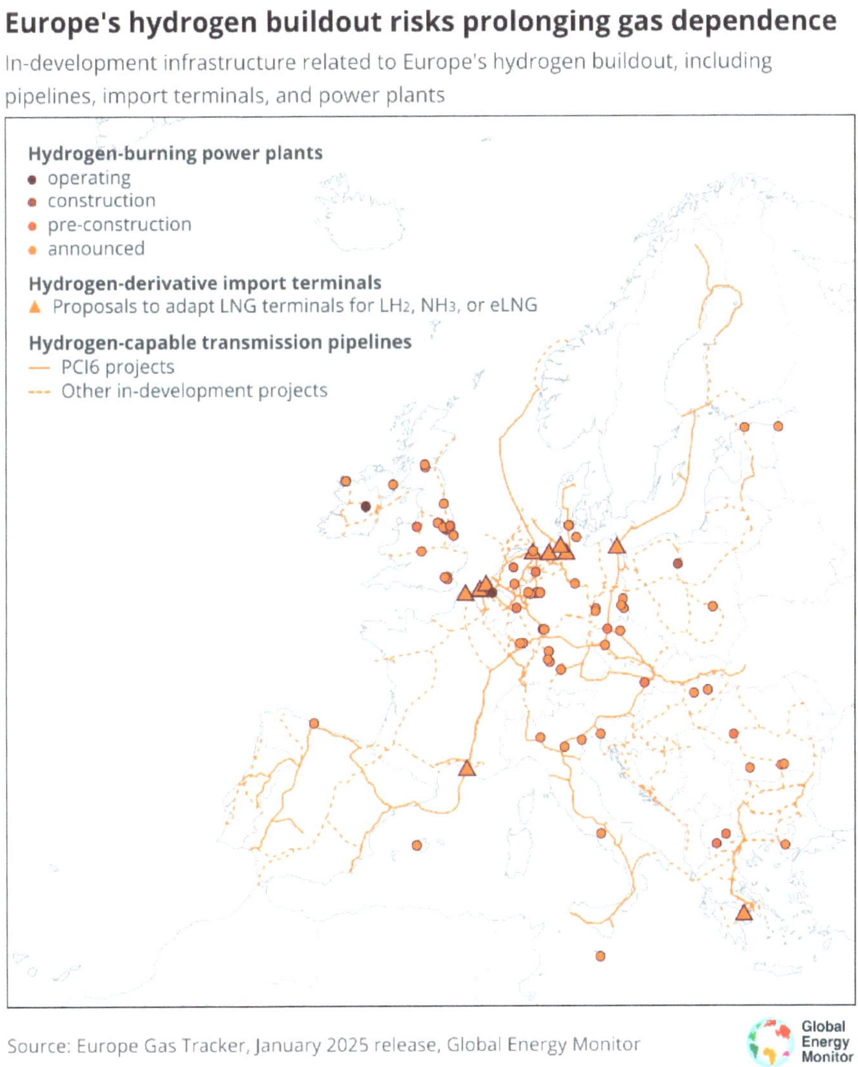

▲ Figure 35: Hydrogen Piping network in Europe

India has an ambitious plan to produce green hydrogen. It has set a target to achieve 5 million tonnes of production by 2030. Setting up solar and wind power and manufacturing of electrolysers are going on a big way. At present, only refineries and some chemical industries have hydrogen piping networks. Industries like IOCL and GAIL are supplying hydrogen blended with natural gas through the existing natural gas pipeline network. India does not have a hydrogen pipeline network for public utility, as of now.

9.6 Hydrogen Transport Through Road

Hydrogen can be transported by road in liquid tanker trucks or tube trailers that carry compressed gaseous hydrogen. Liquid tankers are more economical for long distances due to their larger capacity, while tube trailers are suitable for shorter distances and high-pressure gaseous hydrogen distribution.

Liquid hydrogen transport is favoured for longer distances because a liquid tanker can carry more hydrogen by weight than a tube trailer. Since liquid hydrogen is at -250 ºC, an inbuilt refrigeration system shall be needed, and therefore trucks and tankers are built for this specific purpose. Tube trailers transport gaseous hydrogen, which has been compressed to high pressures, typically 180 bar or higher.

Hydrogen is highly flammable and easily diffuse leading to leakages. Therefore, it requires careful handling and safety measures during transportation. Optimisation of the hydrogen transportation pathway is carried out before dispatch by road to avoid societal risks. Models are developed to identify high-risk paths and optimise transport routes, considering factors like population, environment, and economic losses.

9.7 Hydrogen Transport Through a Ship

Liquid hydrogen can be transported by ship, but it requires specialised tankers and extreme cryogenic temperatures (-253 ºC) due to hydrogen's low density and boiling point. Several projects are underway to develop and demonstrate the feasibility of shipping liquid hydrogen, including the Suiso Frontier, the first liquid hydrogen carrier to transport from Australia to Japan. Kawasaki Heavy Industries is developing a large-scale liquid hydrogen carrier powered by a dual-fuel engine that can run on hydrogen.

Several other companies and research institutions are exploring various aspects of liquid hydrogen shipping, including containment systems, loading/unloading technologies, and ship design.

Challenges in ship transportation are:

- Liquefaction - Hydrogen gas has a very low energy density by volume. To make it transportable by ship, it needs to be liquefied by cooling it to -253°C, which requires significant energy and specialised storage.
- Storage: Liquefied hydrogen (LH_2) needs to be stored in well-insulated tanks to maintain its low temperature. This adds to the complexity and cost of the ships, in addition to boil-off loss.
- Safety: Hydrogen is highly flammable, diffusive, and can embrittle steel, requiring careful design and safety measures to prevent leaks and ensure structural integrity. A boil-off facility to avoid overpressure in the containment is needed.
- Cost: The technology for liquefying, storing, and transporting liquid hydrogen is currently expensive, making it a barrier to widespread adoption.

9.8 Liquefaction of Hydrogen

Hydrogen liquefaction is the process of cooling gaseous hydrogen to extremely low temperatures (below its boiling point, minus 253 degrees Celsius) to bring it into a liquid state, primarily for easier storage and transportation. This process requires significant energy input to reach the necessary cryogenic temperatures and is therefore an energy-intensive and costly endeavour.

How the Hydrogen Liquefaction System works:

- Pre-cooling: Gaseous hydrogen is first pre-cooled to reduce the temperature difference for the subsequent cooling stages.
- Refrigeration: A refrigeration cycle, often involving helium or hydrogen turboexpanders, is used to further cool the hydrogen to near its liquefaction point. This is carried out in stages.
- Liquefaction: The cooled hydrogen is then passed through a Joule-Thomson valve, causing it to expand and cool further, ultimately reaching its liquid state.
- Separation and Storage: The liquid and gaseous hydrogen are separated, and the liquid is stored in cryogenic tanks.

A process flow diagram of the hydrogen liquefaction cycle is shown in Figure 36. It can be seen that cooling takes place in 4 phases: pre-cooling, cryogenic cooling, liquefaction, and finally supercooling to the level of minus 253 degrees Centigrade. Mixed refrigerants (MR) are used in different stages.

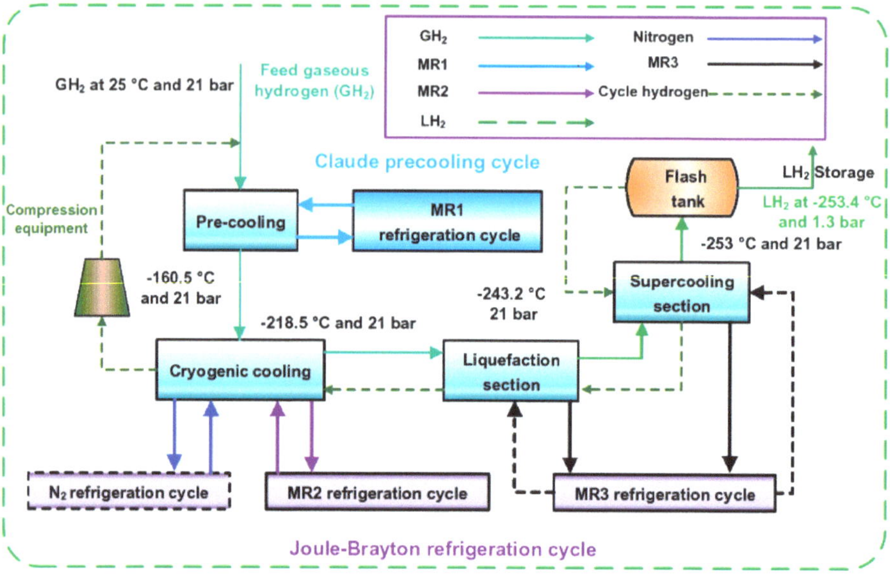

▲ Figure 36: MR-Mixed Refrigerant

Economy and advantages of liquid hydrogen transportation through a ship.

Liquid hydrogen transport by ship presents a technically feasible pathway for international hydrogen trade with the benefit of reduced emissions and potential for decarbonising the maritime sector. However, the economic viability needs careful consideration due to high liquefaction costs, energy losses during transport (boil-off), and significant infrastructure investments. Compression and liquefaction alone consume 7 kWh/kg of hydrogen, which is 21% of its usable energy relative to the lower heating value of 33.3 kWh/kg. It has been reported that about one-third of usable energy is consumed in ship transportation, including compression, liquefaction, boil-off, transport, and loading/unloading.

However, there are many advantages to ship transportation, such as

- Higher Energy Density: Liquid hydrogen has a higher energy density compared to gaseous hydrogen, meaning more energy can be transported in smaller volumes, which is crucial for shipping efficiency.
- Suitability for Long-Distance Transport: Liquid hydrogen is well-suited for long-distance international trade routes, connecting regions with abundant renewable energy resources to energy-consuming industrial centres.

- Established Infrastructure: While a separate infrastructure is needed for hydrogen liquefaction and regasification, leveraging existing ammonia or LNG shipping infrastructure, where possible, could streamline the transition to LH_2 transport.

Ongoing research and development aim to address the challenges of high-cost boil-off and the safety aspects involved, to improve the overall efficiency and cost-effectiveness of liquid hydrogen transport via ships, and ultimately determine its role in the global hydrogen economy.

10 Hydrogen Fuel Cell (HFC)

An HFC is a device that produces electricity through an electrochemical reaction between hydrogen and oxygen, producing water and heat as byproducts. It's a clean energy technology used in various applications, including transportation, stationary power, and portable devices.

10.1 Working Principle of Hydrogen Fuel Cell

Electrochemical Reaction: Hydrogen and oxygen electrochemically react to form water and generate electricity. The Gibbs Free Energy change for this electrochemical reaction is negative (-237 kJ/Mol), so it is spontaneous: no external energy is required to initiate the reaction. Electrical energy, equivalent to Gibbs Free Energy, is produced theoretically when hydrogen combines with oxygen to make water in an HFC. The thermogenic way (combustion) of combining hydrogen and oxygen will produce thermal energy in place of electrical energy in an HFC.

- Components: It is an electrochemical galvanic cell that consists of an anode, a cathode, an electrolyte, and a conductor, similar to an electrolytic cell. An electrochemical cell, like a solid oxide electrochemical cell, can work both as an electrolytic cell and as a galvanic cell.
- Hydrogen Input: Hydrogen gas is supplied at the anode, where a catalyst (often platinum) splits the hydrogen molecule to atomic hydrogen and then atomic hydrogen to hydrogen positive ions (protons) and electrons.
- Protons and Electron Separation: The protons pass through the electrolyte, a PEM, that allows only protons to pass through, while the electrons are directed to flow through a conductor outside. Both combine at the cathode to form water in combination with oxygen from the air.
- Electrochemical reactions are:

Hydrogen molecules are split into protons (H+) and electrons (e-) at the anode.

$2H_2 \rightarrow 4H^+ + 4e^-$

Protons, electrons, and oxygen combine to form water at the cathode.

$$O_2 + 4H^+ + 4e- \rightarrow 2H_2O$$

The function of an HFC is just the opposite of that of an electrolyser. While an electrolyser transforms electrical energy into chemical energy in the form of hydrogen, an HFC transforms chemical energy, in the form of hydrogen, back to electrical energy to drive wheels.

The electrolyser and fuel cell characteristics are schematically shown in Figure 37. Standard electrochemical potential difference between anode and cathode of a fuel cell at standard pressure 1 bar and standard temperature 298 K is 1.23 volts.

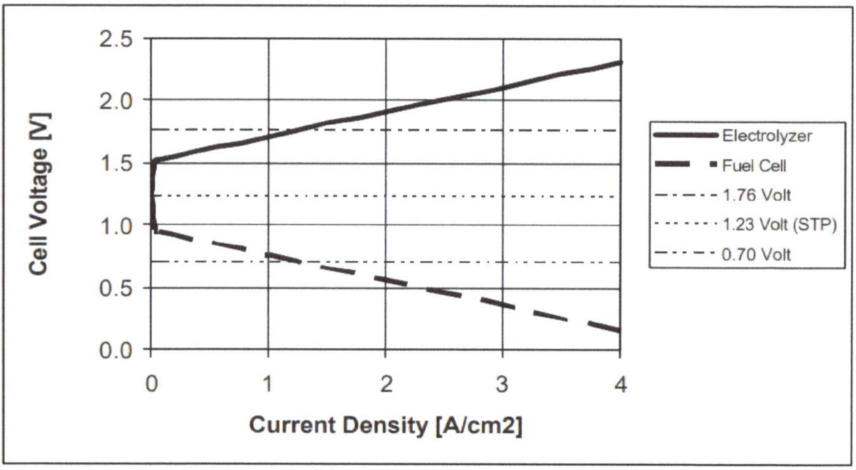

▲ Figure 37: Voltage-current characteristics of the electrolyser and the fuel cell

Assuming that the same electrolyte and catalysts are used in the electrolyser and HFC, the polarisation losses are typically 0.28 volts for solid polymer or alkaline systems. The apparent open-circuit voltages thus become 0.95 volts for the fuel cell and 1.51 volts for the electrolyser, respectively. 0.28 V is reduced from 1.23 in a fuel cell, which is a galvanic cell as we draw current from, whereas 0.28 V is added to 1.23 in an electrolytic cell, where we supply current to. For both, we assume an area-specific resistance of 0.2 W cm^2 and construct the characteristics of a low-temperature fuel cell (dashed line) and a corresponding electrolyser (solid line). Fuel cells are normally operated at 0.7 volts to optimise the system efficiency. We assume the same optimisation requirements apply to an electrolyser. In this case, the corresponding voltage of operation is 1.76 volts as indicated by the dash-dot lines.

Voltage loss is caused by resistance due to cell polarisation and is a measure of electrical losses in a functioning electrolyser or galvanic cell. The losses depend on the current density (higher current density corresponds to a higher rate of H_2 production) in the case of an electrolyser, and on the current drawn from the fuel cell to the motor in the case of a fuel cell.

Figure 38 shows the configuration of an HFC.

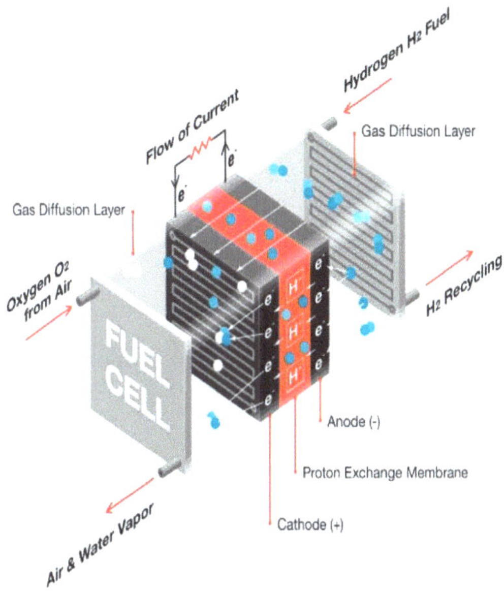

▲ Figure 38: Hydrogen fuel cell technology

10.2 Key Advantages of a hydrogen fuel cell are

Clean Energy: Fuel cells produce electricity without any carbon emissions. Only steam is emitted in hydrogen combustion.

- Versatile Applications: They can be used in a wide range of applications, from powering vehicles to providing backup power for buildings.
- High Energy Efficiency: Fuel cells can be more energy efficient than traditional combustion engines. The energy efficiency of an HFC is around 48%, compared with 25-28% for an internal combustion engine. The life and efficiency of a fuel cell depend mainly on the extent of purity of hydrogen. Hydrogen is used in liquid form to ensure maximum purity. Any contamination shall spoil the catalyst, hence life. HFC comes next to the battery in terms of energy efficiency. A battery is about 80% energy-efficient, since losses occur only during charging/discharging.

Hydrogen fuel cell stacks typically last for the lifetime of a vehicle, roughly 150,000 to 200,000 miles (about 5,000–10,000 operating hours) for passenger cars, and up to 20,000–30,000 hours for heavy-duty trucks and buses. Performance degrades over time due to component wear, but they are designed for long-term, durable operation, often comparable to internal combustion engines.

Current research aims for 80,000+ hours for stationary power, with new catalysts potentially enabling over 100,000 hours to exceed diesel engine longevity.

Figure 39 shows the energy loss (or efficiency loss) at different stages when renewable power is used to drive the wheel via different pathways.

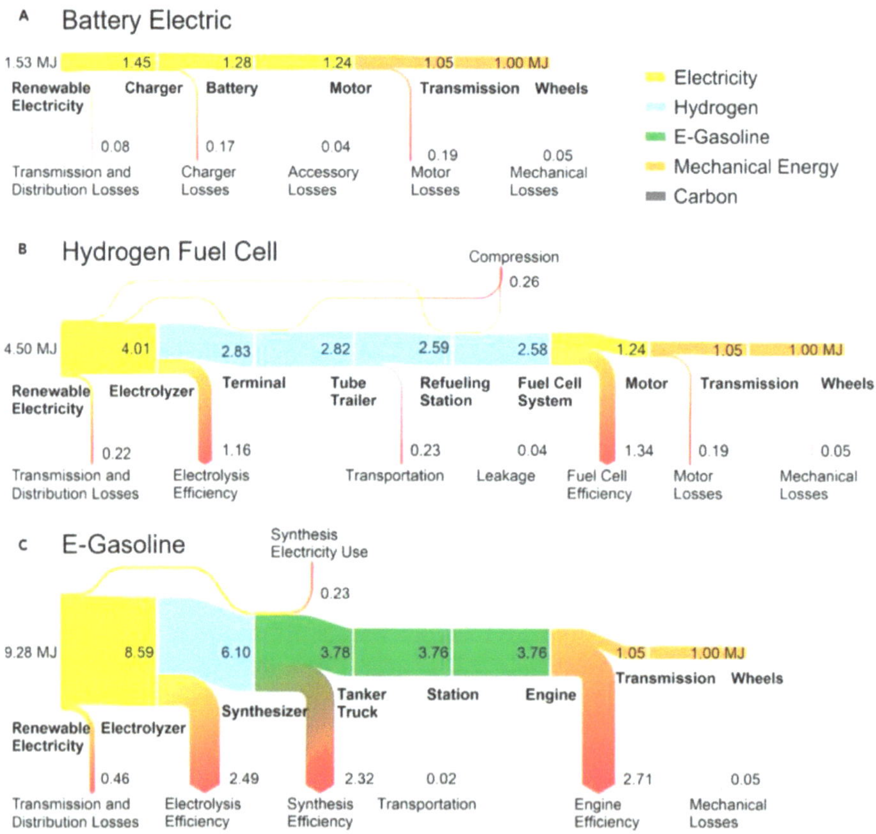

▲ Figure 39: Losses while renewable electricity is fed to the wheel in different pathways

10.3 Challenges in Fuel Cell Technology

Cost: Fuel cells can be more expensive than other energy technologies.

- Hydrogen Infrastructure: The availability of hydrogen refuelling stations is still limited in many areas.
- Pure and Green Hydrogen Supply: Only electrolytic and liquified hydrogen is suitable because of its higher purity requirement for catalysis. Even a slight contamination with H_2 can cause a significant loss of catalytic activity. Purity requirement is as high as 99.999%.
- Durability and Maintenance: Fuel cells have a limited lifespan, less than diesel engine, and require proper maintenance due to wear and tear arising from ionic transport.

The arrangement of the different devices in a hydrogen fuel cell-driven car is shown in Figure 40.

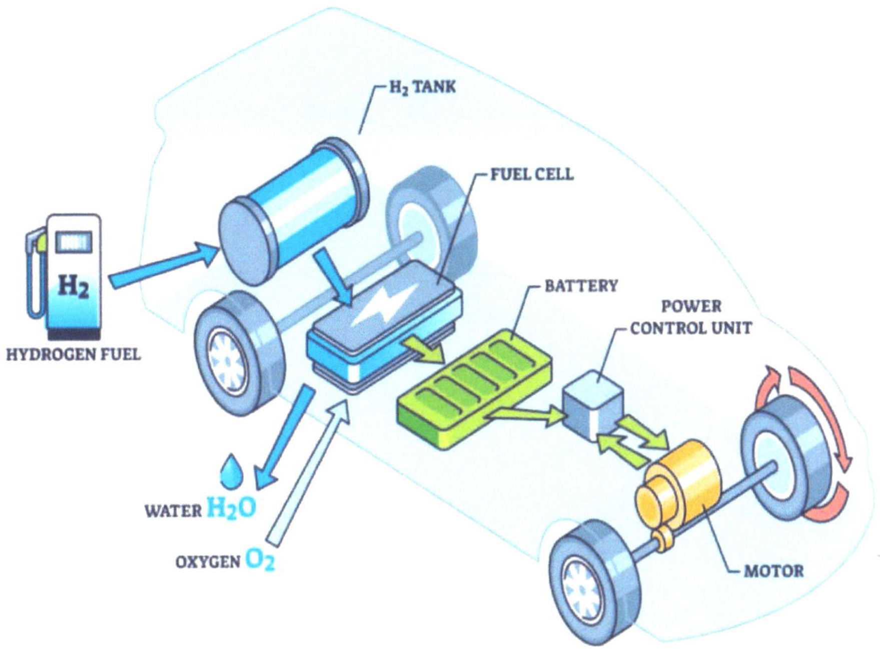

▲ Figure 40: Hydrogen Fuel Cell Car

11 Hydrogen in Industrial Decarbonisation

Apart from fuel, hydrogen is used as the feedstock for the production of many chemicals. Oil refineries, ammonia, and methanol-manufacturing industries use hydrogen the most. Currently, they use grey hydrogen derived from fossil. However, encouragement, incentives, and green regulations by governments and industries are moving towards clean hydrogen in a big way. It is estimated that by mid-thirties, clean hydrogen shall supersede grey hydrogen in terms of global demand because of the reduced cost of production, and by the middle of the century, 70% of hydrogen consumption, globally, shall be clean hydrogen, green or blue. Many agencies, including McKinsey, have predicted this to be as high as 95%.

The following industries consume hydrogen in a big way.

11.1 Oil Refineries

Hydrogen is widely used in oil refining processes. It plays a crucial role in reducing impurities, mostly sulphur in crude oil, enabling the production of cleaner-burning gasoline and diesel fuels.

Hydrogen is used in hydrotreating and hydrocracking processes in refineries to improve the quality and yield of refined utility products. The hydrotreating process removes sulphur from crude and refined products by absorbing sulphur to form hydrogen sulphide.

In hydrocracking, hydrogen's role is crucial. Hydrocracking breaks large hydrocarbon molecules into smaller ones. It is used to produce valuable products (such as gasoline and jet fuel) while simultaneously saturating unstable cracked products, removing impurities (sulphur and nitrogen), and preventing catalyst deactivation by inhibiting coke formation, leading to stable, clean fuels.

Alkanes (saturated hydrocarbons) and alkenes (unsaturated hydrocarbons) are produced through thermal cracking, which is fed to petrochemical industries for high-value products. During the thermal cracking process, hydrogen is produced and used internally by refineries to improve the quality of products.

Hydrogen is a primary feedstock for many organic and inorganic chemicals, including methanol (CH3OH), ammonia (NH3), hydrochloric acid (HCl), hydrogen peroxide (H_2O_2), and many others.

It is used to meet the increasingly stringent product specifications and environmental regulations for fuel production.

As refineries face pressure to reduce their carbon footprint. The use of green hydrogen is becoming increasingly important. Indian refineries currently consume approximately 3 million tonnes per annum (MMTPA) of hydrogen, primarily for desulfurisation and upgrading to higher-value products. This accounts for nearly half of India's total hydrogen consumption.

The vast majority of this hydrogen demand is currently met through domestic production, primarily using carbon-intensive methods like SMR from natural gas. Of the total 5.9 million tonnes of annual requirement, 62% is produced from natural gas, 22% from oil by refineries as a byproduct, which is consumed internally, and 18% from coal.

Industries are planning to gradually increase the share of green hydrogen in the industrial sector, including refineries, to 25% by replacing grey hydrogen. This indicates a strong push to reduce the country's dependence on imported fossil fuels for hydrogen production and to promote cleaner energy alternatives.

11.2 Ammonia and Methanol Industries

Hydrogen is the basic feedstock for ammonia production. Nitrogen and hydrogen are synthesised to produce ammonia, which serves as a feedstock for the urea plant.

Indian fertiliser industries consume a significant amount of hydrogen, primarily in the form of grey hydrogen derived from fossil fuels. In India, grey hydrogen consumption in the fertiliser and refining industries together is estimated at around 6 million metric tonnes per annum (Mtpa).

According to a report by the Council of Energy, Environment, and Water (CEEW), dated 18 June 2024, hydrogen consumption in the fertiliser and refining industries could nearly double to 10.6 Mtpa by 2030, in line with the respective sectoral production targets. The Ministry of Chemicals and Fertilisers aims to increase domestic production to offset fertiliser imports, as outlined. For the refining industry, the Ministry of Petroleum and Natural Gas (MoPNG) aims to increase capacity by setting up new refineries across the country. Similar increases can be expected in the petrochemicals sector, where consumption may grow to 1.3 Mtpa for the same reasons. Together, these

3 sectors are expected to drive India's total hydrogen demand to 11.9 MMTA in 2030.

Methanol is a chemical building block for hundreds of products that touch our daily lives. It is used in everyday products such as paints, carpeting, plastics, and more. Increasingly, methanol is being employed as an emerging energy product around the world. It is used to fuel cars and trucks, marine vessels, boilers, cookstoves, and kilns. In addition to its growing list of market applications, methanol is also used as a superior hydrogen carrier. As one of the most versatile chemical commodities and energy sources available today, methanol facilitates the energy transition of economies towards a greener and more sustainable future. It also represents a pathway that enables the development of hydrogen economies. With the highest hydrogen-to-carbon ratio of any liquid fuel, methanol can be highly instrumental in realising hydrogen adoption. As a liquid, it circumvents the common challenges associated with the logistics required for the transportation of hydrogen. It is a globally traded commodity with a global production capacity of 148 million metric tonnes (MMT) to meet a global demand of nearly 100 MMT. There is a mature global supply chain that can be leveraged to store, transport, and distribute methanol as a hydrogen carrier. These supply chains exist in India today and can readily support the delivery of methanol to end users who will reform it into hydrogen for their necessary applications.

Hydrogen is the feedstock for methanol production. It is produced by hydrogenating carbon dioxide. Synthesis gas, a mixture of H_2 and carbon monoxide, is also used to manufacture methanol through a catalytic process.

Globally, 100 MMTs of methanol was produced and consumed in 2020. India has a methanol production capacity of approximately 1MMT. This number is derived from the combined production capacity of methanol plants owned by Assam Petrochemicals, Rashtriya Chemicals and Fertilisers, and Gujarat State Fertilisers & Chemicals. The country's methanol demand is projected to grow to 4.675 MMT by 2030. It can be produced from diverse feedstocks. Conventionally, methanol is produced from coal and natural gas. Increasingly, sustainable feedstocks such as captured carbon dioxide, municipal solid waste, and agricultural waste are also used to produce renewable methanol. The adoption of methanol as a fuel or hydrogen carrier provides the industrial and transport sectors with a roadmap to achieve net carbon neutrality.

Methanol is an excellent hydrogen carrier, packing more hydrogen in this simple alcohol molecule than can be found in hydrogen that has been compressed (350–700 bar) or liquified (-253˚C).

By 2030, the primary sectors driving hydrogen demand in India are projected to be refineries and fertilisers, with additional growth in steel and chemicals, though specific demand figures vary across reports. While overall domestic hydrogen demand is estimated at around 11 million tonnes per annum by 2030, the NGHM aims to achieve a significant share of this as green hydrogen, potentially reaching 5 MMT out of the total 12 MMT demand.

Figure 41 below indicates total hydrogen demand and share of hydrogen by 2050 in India, by major hydrogen consumer industries, refineries, fertilisers, and petrochemicals, taken from RMI analysis.

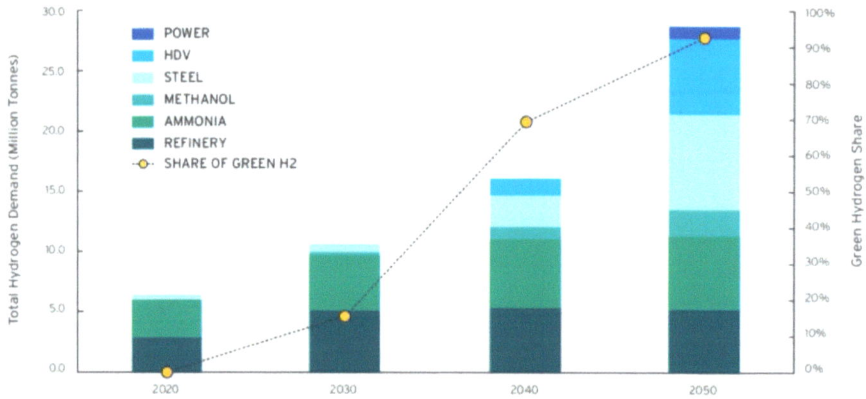

Source: MoS, MoC&F, MoPNG, IEA, TERI, BCG, World Bank, RMI Analysis

▲ Figure 41: Total hydrogen demand and share of green hydrogen by 2050 in India

11.3 Steel and Cement Industries

The steel industry is a significant emitter of carbon dioxide, contributing roughly 8% of global CO_2 emissions. Decarbonising this sector is crucial for achieving global climate goals and requires a multi-faceted approach. Key strategies to decarbonise steel industries include transitioning to low-carbon production methods, increasing scrap recycling, and implementing carbon capture technologies.

Presently, the dominant steelmaking process, the blast furnace and basic oxygen furnace (BF-BOF) route, relies on coal, which is highly carbon-intensive.

Decarbonisation Strategies for Steel Industries are through:

- **Electric Arc Furnaces (EAFs):** EAFs, powered by renewable electricity, offer a pathway to near-zero emissions steel production, particularly when recycling the scrap steel.
- Hydrogen-based steelmaking**:** Replacing coal with green hydrogen in DRI production is a promising avenue for decarbonisation. In the Hydrogen-based Direct Reduced Iron (H_2-DRI) process, hydrogen acts as a reducing agent, reacting with oxygen in iron ore to produce metallic iron and water. This process replaces traditional methods that rely on coal or natural gas, which produce carbon dioxide (CO_2) as a byproduct. The H_2-DRI process is considered environmentally friendly due to its lower carbon emissions.

In the DRI process, the reduction reaction can be represented as:

$$Fe_2O_3 + 3H_2 \rightarrow 2Fe + 3H_2O \text{ or}$$
$$Fe_3O_4 + 4H_2 \rightarrow 3Fe + 4H_2O$$

The reaction takes place at around 1,000°C, well below the melting point of iron. The iron produced is called sponge iron (when produced in a blast furnace, it is called pig iron), and its iron content is above 90%. Sponge iron is fed into the steel industry, similar to scrap iron.

The H_2-DRI process is a key component in the transition towards more sustainable steelmaking, as it offers a pathway to significantly reduce carbon emissions.

The steel industry is one of the 3 biggest industries, along with power and cement, that emit the maximum CO_2. All 3 industries together account for around 50% of global industrial emissions. Every tonne of steel produced in 2018 emitted, on average, 1.85 tonnes of CO_2, amounting to about 8% of global emissions, according to the World Steel Association.

Decarbonisation strategy of cement industries**.**

The decarbonisation of the cement industry is a significant challenge, primarily due to the inherent CO_2 emissions from the clinker production process (calcination of limestone) and its high energy consumption. Many strategies are being explored and implemented to reduce emissions, including innovative materials, carbon capture technologies, alternative fuels, and energy efficiency improvements.

Clinker is a nodular material, essentially a precursor to cement. Clinker is produced by heating a mixture of limestone and other materials, such as clay, at high temperatures in a kiln, causing a chemical reaction that forms hard, glassy nodules. These nodules are then ground and mixed with other ingredients, such as gypsum, to produce the final cement product. The CO_2 evolution reaction called calcination is as mentioned below:

$$CaCO_3 = CaO + CO_2$$
$$2CaO + SiO_2 = Ca_2SiO_4 \text{ and}$$
$$Ca_2SiO_4 + CaO = Ca_3SiO_5$$

Ca_3SiO_5, called clinker, gives cement its binding strength. This process takes place at very high temperatures above 1,300 degrees Centigrade, which requires large amounts of fossil fuel to burn, and, at the same time, CO_2 is produced as a byproduct.

Since carbon dioxide is a byproduct in clinker, CCUS technology can make it clean until an alternative to clinker as a binding material is invented for the cement industry.

The cost of decarbonising steel and cement is shown in Figure 42, taken from the IEA report.

BF-BOF – Blast Furnace- B

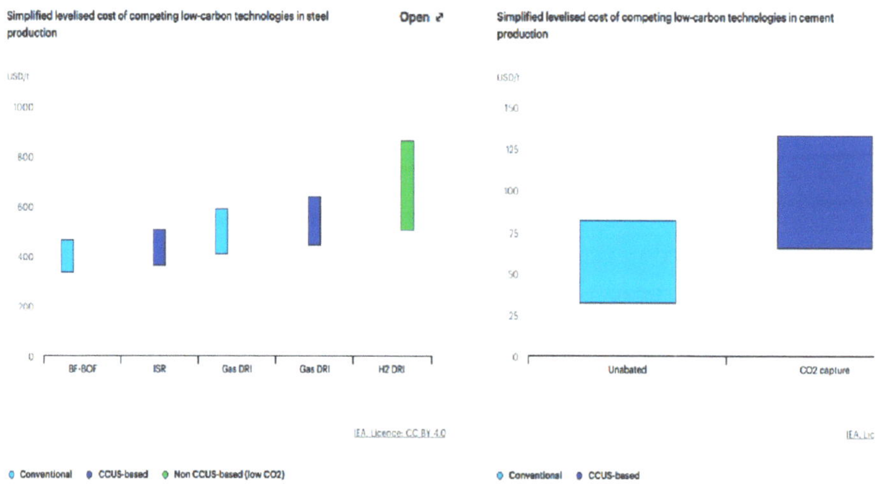

(BF-BOF – Blast Furnace-Basic Oxygen Furnace, ISR- Incremental Shape Rolling, Gas DRI – Natural Gas Direct Reduced Iron.)

▲ Figure 42: Decarbonising the cost of the steel and cement industries

11.4 HEFA and F-T Process for Sustainable Aviation Fuel (SAF) and Synthetic Fuel Production.

A global energy transition is accelerating to address climate change through innovative decarbonisation and carbon capture and utilisation (CCU) technologies. These efforts aim to establish renewable energy and circular carbon systems, transforming CO_2 into high-value materials and synthetic fuels to achieve a closed carbon loop. Simultaneously, the oil, gas, and mobility sectors are integrating electrification and AI-driven solutions, such as digital twin technologies, to enhance efficiency and reduce emissions. The aviation industry is advancing towards sustainable aviation fuels (SAFs) using catalytic routes such as HEFA (Hydro processing of Ester and Acetic Acids) from diverse bio-based feedstocks and Fischer-Tropsch (F-T) synthesis. Both HEFA and FT processes are based on green hydrogen feedstock to make the product green.

The FT process uses syngas, consisting of H_2 and CO, as a feedstock. The syngas is passed through a selective catalytic reactor, and the product is separated in fractional distillation to separate out SAF-grade fuel.

The HEFA process uses renewable lipids, specifically waste oils, fats, and greases, alongside vegetable oils, as feedstock for SAF production. These include used cooking oil (UCO), tallow (animal fat), and refined vegetable oils.

A process flow diagram of the HEFA process using H_2 is shown in Figure 43.

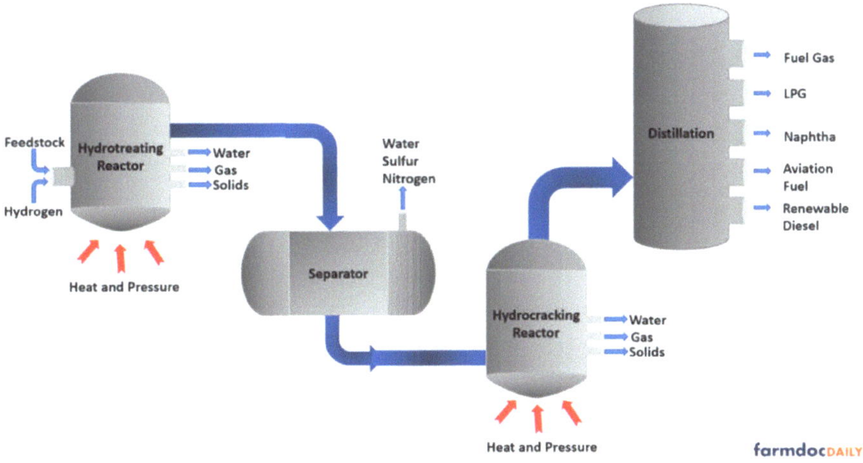

▲ Figure 43: HEFA Sustainable Aviation Fuel (SAF) production process

Catalytic technology will play a crucial role in making the green hydrogen and synthetic fuel production process energy-efficient and economical. Efficient electrolysers, such as PEMs, use very costly platinum group metal catalysts, such as iridium, platinum, and palladium, which make the produced green hydrogen expensive. Finding an efficiently active and economically affordable catalyst makes the process energy-efficient and cost-effective, so that it can compete with fossil fuels in the energy market. Producing green fuel requires energy-efficient electrolysers, cost-effective carbon capture, and carbon recycling through synthetic fuel, and all these can be addressed by advancements in catalytic technology and upgrading process operations. The commercial success of all green fuel production processes, such as electrolysis, FT, ATR, HEFA, ATJ (alcohol-to-jet fuel), and CO_2 hydrogenation, largely depends on effective catalysis.

11.5 Green Hydrogen Demand in India

According to Niti Aayog's estimates, around 24% of the industrial sector's hydrogen demand in FY 2030 can be met by green hydrogen. India's projected total green hydrogen demand in 2030 shall be approximately 1.1 Mtpa. Projected green hydrogen demand in India by industrial sectors, from FY 2030 to FY 2050 (Mtpa), is given in Figure 44.

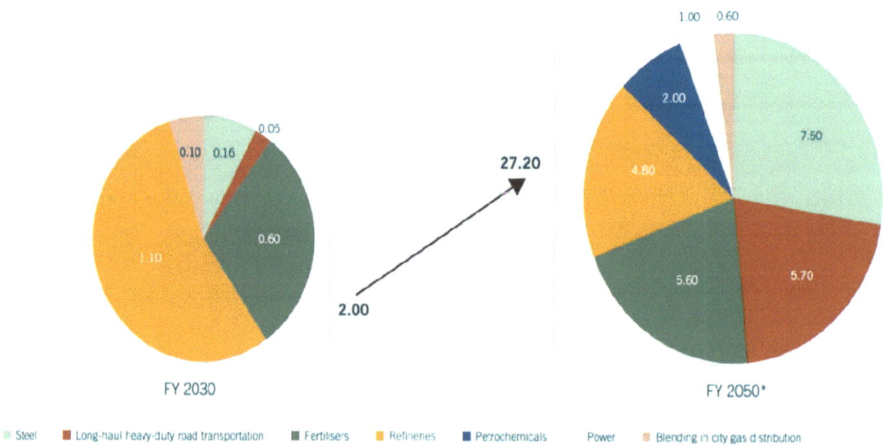

Source: CEEW analysis

▲ Figure 44: Green hydrogen demand by Niti Aayog

Council on Energy, Environment and Water (CEEW), an independent, non-partisan policy research institution and think tank based in India, focused on sustainable resource use, energy, climate, and water issues, in India and globally, has highlighted the green hydrogen prospects in its report, dated 18 June 2024, on *Financing Green Hydrogen in India*. The salient points of the report are:

- Green hydrogen demand in India is projected to exceed 27.2 million tonnes per annum by 2050, primarily led by non-energy sectors such as steel, fertilisers, refineries, and long-haul heavy-duty road transportation applications.
- By leveraging its competitive advantage in renewable energy costs, India can emerge as a leading exporter of green hydrogen, especially in the short-to-medium term.
- The cost of switching from grey to green hydrogen remains the largest impediment to the industrial uptake of green hydrogen in India. According to current industry estimates, green hydrogen production costs range from USD 3.5–5 per kg, while breakeven costs for green hydrogen to replace conventional fuels are below USD 1.5 per kg.
- A nascent enabling infrastructure for green hydrogen deployment and high financing costs are other prominent challenges faced by the sector.
- Policy considerations such as waiver of power banking and open access charges, reducing the GST on electrolysers and renewable energy components, availing incentives under the Strategic Interventions for Green Hydrogen Transition (SIGHT) scheme of Indian government, accessing cheaper green debt for financing green hydrogen projects coupled with market-driven innovations in electrolyser and renewable energy technologies, can cumulatively reduce the levelised cost of green hydrogen by around USD 1.8 per kg in India.

12 National Green Hydrogen Mission (NGHM) of India

12.1 Background of NGHM

India has declared its goal to achieve net-zero emissions by 2070. As India's growth story unfolds, its demand for energy and resources is set to rise. Energy use has doubled in the last 20 years and is likely to grow by at least another 25% by 2030. India currently imports over 40% of its primary energy requirements, worth over USD 90 billion every year. Major sectors such as mobility and industrial production are heavily dependent on imported fossil fuels. This necessitates a shift towards technologies that increase the share of renewable energy in the energy mix and progressively reduce reliance on fossil fuels. India holds as low as only 0.3% of global oil reserves and 0.6% of natural gas, but as high as 18% of the global population, with only 2.4% of the land, and only 4.2% of global potable water reserves. Therefore, India needs to move towards renewable energy to ensure energy security.

Green hydrogen has the potential to play a key role in such low-carbon and self-reliant economic pathways. Green hydrogen can enable the utilisation of domestically abundant renewable energy resources across regions, seasons, and sectors, feeding multiple usage streams, either as a fuel or as an industrial feedstock. It can directly replace fossil-fuel-derived feedstocks in petroleum refining, fertiliser production, and steel manufacturing. Hydrogen or its derivatives, which fuel long-haul automobiles and marine vessels, can enable decarbonisation of the mobility sector. Green hydrogen can be particularly useful as a versatile energy carrier for meeting the energy requirements of remote geographies, including islands, in a sustainable manner, as is being done with LPG today.

Many major economies have adopted hydrogen strategies as part of broader climate and clean energy initiatives. Following the energy crisis in Europe, which intensified due to the Russia-Ukraine war, European nations are moving to replace natural gas with hydrogen. These national strategies largely aim to address common challenges in scaling up green hydrogen production, enhancing hydrogen use across sectors, developing technologies, and designing enabling policies and regulations.

As global consensus on net-zero gathers momentum, demand for green hydrogen and its derivatives is set to rise. The asymmetries in expected demand and production capabilities for green hydrogen in different countries and regions are likely to result in international trade of green hydrogen and its derivatives, like green ammonia and green methanol. The sensitivity of fossil fuels to geopolitical upheavals and the experience of supply chain disruptions during COVID have accelerated the transition towards green fuels and feedstocks. This presents a unique opportunity for India to capitalise on its abundant renewable energy and land resources, and on the growing global demand for green hydrogen, to become a leading producer and exporter of green hydrogen and its derivatives.

Despite the unique possibilities and advantages, unfavourable cost economics, lack of harmonised standards and regulations, supply challenges, and costly enabling infrastructure have thus far held back the replacement of fossil fuels and fossil fuel-based feedstock with green hydrogen or its derivatives. However, recent trends and analyses indicate that, driven by technological advancements, declining costs of renewable energy and electrolysers, and aggressive national strategies by some major economies, green hydrogen is likely to become cost-competitive across industry, mobility, and other sectors within a short span.

Rapid deployment of renewable energy and electrolyser capacities will be required to achieve economies of scale. Associated infrastructure and regulatory ecosystem will need to be established to deliver renewable power and to support the storage, transportation, and utilisation of green hydrogen for various applications. Accelerated technology development to improve performance, efficiencies, safety, and reliability would also be crucial.

The green hydrogen pathway can be a key enabler for India's aspirations of building a low-carbon and self-reliant economy.

The Government of India, therefore, decided to set up and implement the National Hydrogen Mission through a comprehensive, integrated approach involving various Central and State Government agencies. The MNRE was entrusted with overall coordination of the Mission's implementation.

With the above background, the Prime Minister of India launched the NGHM on the 75th Independence Day, 15 August 2021, and formalised it in January 2023. NGHM aims to establish the country as a global hub for the production, usage, and export of clean hydrogen by 2030.

12.2 Objective of the Mission

The objectives of NGHM are

1. The objective of the Mission is to make India the Global Hub for production, usage, and export of green hydrogen and its derivatives. This will contribute to India's aim to become *Aatmanirbhar* (self-reliant) through clean energy and serve as an inspiration for the global clean energy transition. The Mission will lead to significant decarbonisation of the economy, reduced dependence on fossil fuel imports, and enable India to assume technology and market leadership in green hydrogen.
2. To achieve the above objectives, the Mission will build capabilities to produce at least 5 MMT of green hydrogen per annum by 2030, with potential to reach 10 MMT per annum with the growth of export markets. The Mission will support replacing fossil fuels and fossil fuel-based feedstocks with renewable fuels and green hydrogen-based feedstocks. This will include the replacement of hydrogen produced from fossil fuel sources with green hydrogen in ammonia production and petroleum refining, blending green hydrogen in city gas distribution systems, production of steel with green hydrogen, and use of green hydrogen-derived synthetic fuels (including green ammonia, green methanol, etc.) to replace fossil fuels in various sectors including mobility, shipping, and aviation. The Mission also aims to make India a leader in the technology and manufacturing of electrolysers and other enabling technologies for green hydrogen.

The expected outcome of the Mission by 2030 is shown in Figure 45.

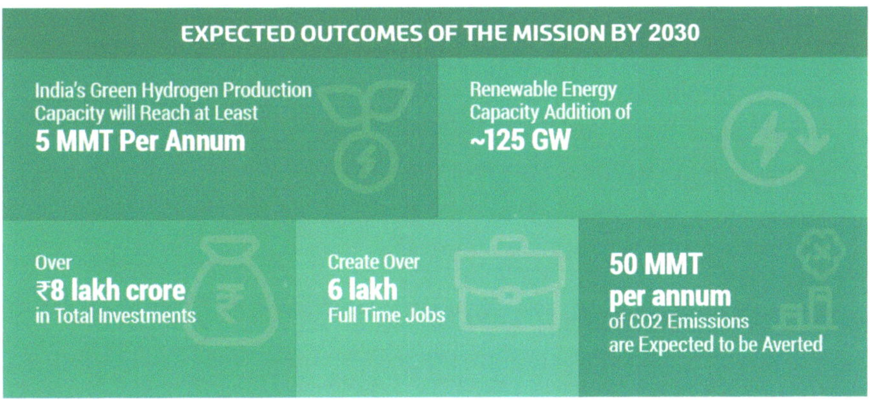

▲ Figure 45: Expected Outcome of NGHM, by 2030

12.3 Initiatives by the GoI in the Implementation of NGHM

1. Financial Incentives: SIGHT Programme
   The GoI has implemented several key initiatives to achieve these goals: The Strategic Interventions for Green Hydrogen Transition (SIGHT) programme provides financial support with an outlay of ₹17,490 crores. This includes incentives for manufacturing electrolysers and producing green hydrogen and green ammonia.
2. Pilot Projects in Hard-to-Abate Sectors
   The mission supports pilot projects to integrate green hydrogen into sectors like shipping, steel, and transport, with an allocation of ₹1,466 crores. This includes projects for green methanol bunkering and the use of fuel cell vehicles. A high-altitude green hydrogen mobility project has also been commissioned in Leh.
3. Development of Green Hydrogen Hubs
   Regions suitable for large-scale production and use have been identified as Green Hydrogen Hubs. Three major ports are recognised as such, with support for developing the necessary infrastructure.
4. Regulatory Framework and Standards
   The Green Hydrogen Certification Scheme of India (GHCI) defines "green" hydrogen based on emission standards. The Ministry of New and Renewable Energy (MNRE) is also developing regulations and standards for safety, storage, and transportation.
5. Enabling Policy Measures
   Policies include waiving Inter-State Transmission System (ISTS) charges for green hydrogen/ammonia plants commissioned before December 31, 2030, for 25 years. Exemptions from prior Environmental Clearance for certain plants and duty benefits under the SEZ Act are also provided.
6. Research, Development, and Skill Generation
   A dedicated R&D scheme with ₹400 crore supports projects in various areas. Initiatives like the Strategic Hydrogen Innovation Partnership (SHIP) and a scheme for startups encourage technological advancement. Skill development programmes are also in place.
7. International Cooperation
   MNRE has collaborated with various countries through MoUs and partnerships, including forming the India-US Hydrogen Task Force.

These initiatives are projected to generate over 6 lakh jobs, attract ₹8 lakh crore in investment, reduce annual CO_2 emissions by nearly 50 MMT by 2030, and create 6 lakh jobs.

India is witnessing a major deployment of large-scale green hydrogen projects, backed by increased electrolyser manufacturing. Domestic companies, such as Ohmium Operations Private Limited, Advait Infratech, and L&T electrolysers, are a few of the key players in this sector contributing to the country's efforts to achieve energy independence and reduce carbon emissions.

Many renowned electrolyser manufacturers all over the globe, including Nel and Ohmium, have shown interest in setting up their factories in India.

Ohmium Operations Private Limited (Ohmium) is a PEM-based electrolyser company. It has a hydrogen electrolyser manufacturing plant in Bengaluru, India, with a 400 MW capacity. It is one of the renowned green hydrogen electrolyser manufacturers in India.

Here are the top 10 green hydrogen electrolyser manufacturers and developers in India:

1. Reliance Industries Limited (RIL): Developing a massive integrated green hydrogen ecosystem in Jamnagar.
2. Larsen & Toubro (L&T) Electrolysers: Partnering with HydrogenPro to manufacture pressurised alkaline electrolysers, with 300 MW capacity planned under the SIGHT scheme.
3. Greenko Zero C (with John Cockerill): Setting up large-scale electrolyser manufacturing (up to 2000 MW/annum).
4. Ohmium Operations India: Launched India's first green hydrogen electrolyser gigafactory in Bengaluru, focusing on modular PEM electrolysers.
5. H2e Power Systems: Developing solid oxide and PEM electrolyser manufacturing, with1000 MW planned capacity.
6. Adani Enterprises (Adani New Industries): Investing in vertical integration for electrolyser production as part of its green energy portfolio.
7. Waaree Energies: A major solar player entering the green hydrogen space, developing electrolyser manufacturing capabilities.
8. Advait Infratech: Approved for 100 MW annual manufacturing capacity under the NGHM.
9. HomiHydrogen: Focused on manufacturing various types of electrolysers for local and export markets.
10. NTPC Ltd: Driving demand by developing large-scale electrolyser installations at its renewable energy parks.

13 Green Hydrogen to Compete with Natural Gas

Green hydrogen is sourced from water electrolysis. It can also be sourced from biomass gasification or biogenic processes, but these processes are more cumbersome due to complex logistics and limited biomass feedstock availability. Therefore, the electrolytic process is the main source of green hydrogen. As discussed earlier in Chapter 4.

For green hydrogen to penetrate the fuel market, it has to compete with natural gas in terms of price, and at the same time, it should be safe to handle. These 2 bottlenecks are discussed separately below.

13.1 Reducing the Cost of the Electrolyser

The Department of Energy Hydrogen Programme Record, dated May 20, 2024, indicates that the modelled levelised cost of green hydrogen, produced from PEM electrolyser and renewable energy sources, is approximately \$5 to \$7 per kilogram (in \$2022, excluding subsidies). These values are based on a range of PEM installed capital costs (average of \$2,000/kW) using various references, real-world data, and analytical models at low manufacturing volumes. Renewable electricity costs of roughly \$0.03/kWh and capacity factors ranging from approximately 50 to 75% are considered.

Table 11 indicates key PEM electrolyser parameters considered in the analysis, representative of recent technology status. The system performance and lifetime parameters, as well as plant parameters including capital and replacement costs, are consistent with an industry peer-review conducted in 2023 and with modelling conducted by the DOE H2NEW Consortium.

▼ Table 11: Considered plant parameters for LCOH calculation

Electrolyzer system electricity usage	~57 kWh/kg
Electrolyzer system lifetime	30 yr
Plant capacity	50,000 kg/day
Installed capital costs	\$1,500; \$2,000; \$2,500/kW [\$2022]
Replacement cost interval	40,000 operating hours
Replacement cost	11% of total installed capital cost

Table 12 shows the levelised costs of hydrogen (in 2022 USD) produced by current PEM electrolyser technology across different scenarios based on electricity sources, with associated inputs for electricity capacity factor and price, and electrolyser capital costs.

▼ Table 12: LCOH at different power supply parameters and electrolyser cost

Scenario[h] *Based on Electricity Source*	Capacity Factor *(%)*	Electricity Price *(¢/kWh)*	Electrolyzer Installed Capital Cost *(2022\$/kW)* \$1,500	\$2,000	\$2,500
			Levelized H_2 Production Cost *(2022\$/kg)*[i]		
Grid – Average Scenario	97%	8.3¢	\$6.80	\$7.50	\$8.20
Renewable Electricity Scenarios					
Hydropower	50%	3.4¢	\$5.50	\$6.70	\$7.90
Land-Based Wind *(Class 1)*	51%	2.9¢	\$5.20	\$6.40	\$7.50
Hybrid Wind-PV	74%	3.3¢	\$4.40	\$5.20	\$6.00

Electrolysers are costly mainly due to the high green energy input (to make hydrogen fully green) required in the manufacturing process, the need for specialised and often costly materials like platinum or other noble metals as catalysts, and the intricate manufacturing processes involved.

The electrolyte used in the PEM electrolyser is a special solid polymer compound that accounts for a significant share of the cost. There is much R&D work going on to reduce the cost of the polymer and improve efficiency. Developing an improved catalyst that breaks water molecules with lower activation energy is also underway; this will increase energy efficiency and lower costs.

Currently, alkaline electrolysers are generally the least expensive at the commercial scale, while PEM electrolysers are more expensive due to the use of precious metals.

Electrolyser manufacturing costs can vary significantly by region, with some regions experiencing higher costs due to factors like inflation and supply chain issues. For example, some reports indicate that electrolysers manufactured in China may be significantly less expensive than those made in Europe or the US. This is because increasing global electrolyser manufacturing capacity is expected to drive down costs through economies of scale. Capital costs for PEM cells, in China, can range from $1,400/kW to $2,500/kW, according to the DOE, and for alkaline cells, from $800 to $1,000 per kW for commercial-scale plants.

A recent report estimates installed electrolyser costs (including equipment, gas treatment, plant balancing, and engineering, procurement, and construction) between $2,000/kW for alkaline and $2,450/kW for PEM electrolysers.

According to the IEA report, the electrolyser cost may fall by 50% by 2030 due to economies of scale.

Analysis of PEM electrolysers indicates that achieving high manufacturing rates (e.g., 1 gigawatt per year) can significantly reduce both the stack and balance-of-plant (BOP) costs. Another report suggests PEM electrolysers could reach around $440– $500/kW in 2030, with economies of scale from multi-GW/year production rates and additional cost reductions from continued R&D-driven technology advances.

13.2 Reducing the cost of green power

The global share of solar electricity is only 5.4%, mainly due to its high capital costs and low efficiency. Even though the running cost of solar power PV cell is negligible, approximately $0.01/kwh, against conventional energy from a thermal power plant $ 0.1/kWh, the capital cost of solar is much higher, which is $4500/kw against conventional thermal power $750/kw. If the chemical process of silicon crystal manufacturing is taken one step further, from polycrystalline cells to monocrystalline cells, which have better electrical conductivity, it becomes further costlier, hence not widely used.

Working Principle of PV Cell and Wind Turbine.

For a better understanding of the photovoltaic cell and its cost components, let us know how a photovoltaic cell works.

A photovoltaic (PV) cell converts light, typically sunlight, into electricity by using the photovoltaic effect: photons from light strike a semiconductor material (like silicon), exciting electrons and creating "electron-hole" pairs. An internal electric field then separates these charge carriers, directing electrons to one side of the cell and holes to the other, generating a potential difference.

The semiconductor material has a built-in electric field, called a depletion layer, often created by a P-N junction. This field acts like a one-way street, separating the newly created electron-hole pairs. Electrons are pushed to one side (the n-type layer) and holes to the other (the p-type layer). This separation of charges creates a voltage across the cell. When an external circuit (like a wire) is connected between the 2 sides, electrons flow through it to recombine with holes, creating a direct current (DC) that can power electrical devices.

Figure 46 shows the working principle of a photovoltaic cell.

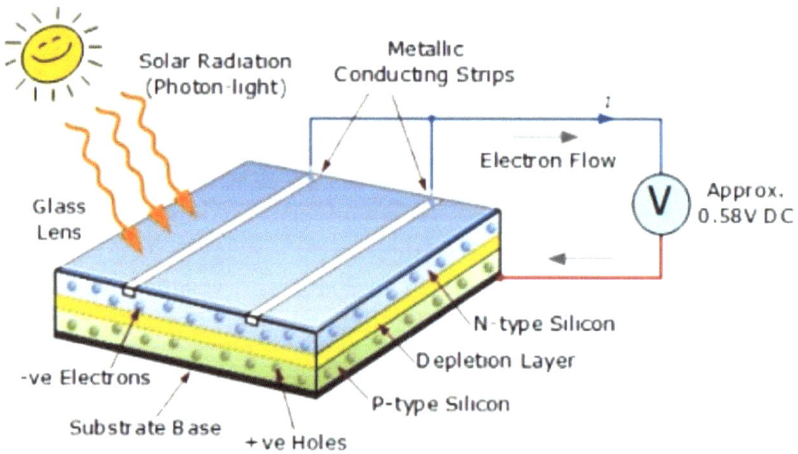

▲ Figure 46: Working principle of PV Cell

Solar cells are categorised by their materials and structures, with common types that include monocrystalline (18–24% efficiency), polycrystalline (13–16% efficiency), and thin-film (7–13% efficiency). Advanced and emerging technologies such as PERC (Passivated Emitter and Rear Cell), N-type HJT (Heterojunction Technology), TOPCon (Tunnel Oxide Passivated Contact), and Perovskite offer higher efficiencies, with some reaching over 25%.

The theoretical maximum energy efficiency of a single-junction solar cell is about 33.7%, known as the Shockley-Queisser limit. Practical, commercially available panels typically convert 15–22% of solar energy into electricity. Factors like poor material quality, temperature, and reflection reduce a panel's real-world efficiency.

The global contribution of solar electricity is 6–7%.

Working Principle of Wind Turbine

Wind turbines work on a simple principle: instead of using electricity to make wind—like a fan—wind turbines use wind to make electricity. Wind turns the propeller-like blades of a turbine around a rotor, which in turn spins a generator that creates electricity. A wind turbine converts the kinetic energy of wind into electricity by using large blades to capture wind, which spins a rotor. This spinning action turns a shaft connected to a gearbox, which increases the rotational speed and drives a generator to produce electrical energy. The generated electricity is then sent to a transformer and into the power grid for distribution.

The fundamental formula for wind turbine power output is obtained from the kinetic energy of moving air masses. The basic equation for calculating wind power available in the wind is as follows,

$$P = \tfrac{1}{2} \times \rho \times A \times V^3$$

Where:

- P = Power (watts)
- ρ = Air density (kg/m^3), taken as 1.225 kg/m^3
- A = Rotor swept area (m^2), which is, 3.14 R^2, R is the length of turbine blade.
- V = Wind speed (m/s), taken as 12 meter/sec, rated wind speed.

The equation illustrates the maximum possible energy that can be derived from the wind, but for wind power as a turbine or wind farm output, you need to factor in turbine efficiency and the Betz limit which no wind turbines can achieve more than 59.3% of the kinetic energy that the wind contains.

Thus, basis the above, the practical formula for wind turbine power output becomes:

$$P = \tfrac{1}{2} \times \rho \times A \times V^3 \times Cp$$

Where Cp is the power coefficient, which shows the turbine's efficiency to convert wind energy to electrical energy, taken as 0.45.

Figure 47 shows the working mechanism of a wind turbine.

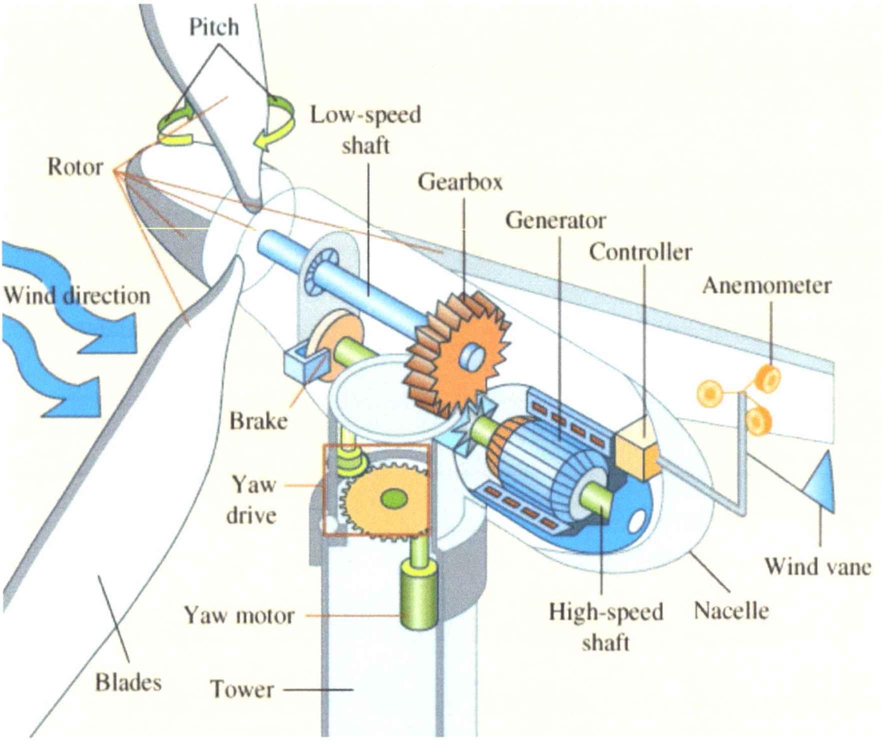

▲ Figure 47: Working mechanism of a wind turbine

According to AleaSoft Energy Forecasting (a pioneer in predictive artificial intelligence), 29 September 2023, over the past decade, the levelised cost of electricity (LCOE) from renewables has dropped sharply. The largest drops in the LCOE of renewable technologies since 2010 have been for onshore wind energy (69%), solar photovoltaic energy (89%), and offshore wind energy (59%). This evolution of renewable LCOE is expected to further drive the adoption of renewable energies.

The LCOE for renewables was calculated by 2 different agencies. One is Lazard's LCOE version 16.0 9 (an AI-based prediction made in 2023), and the other is IRENA's cost estimate up to 2022. There are a few differences between the 2 sources.

IRENA calculates a weighted average LCOE of projects for each technology worldwide. While Lazard's results are more focused on projects in the United States, they are often used as a global benchmark. In either case, the historical trend in the LCOE of renewable technologies is the same.

Figures 48, 49, and 50 indicate the LCOE from solar, onshore wind, and other technologies. It can be seen that the global cost has fallen to as low as \$50/MWh from solar and \$26/MWh from onshore wind, relative to the 2010 cost. LCOE for onshore wind is the lowest among the technologies used for energy production. The cost is 2.5 years old and is expected to come down further in today's terms.

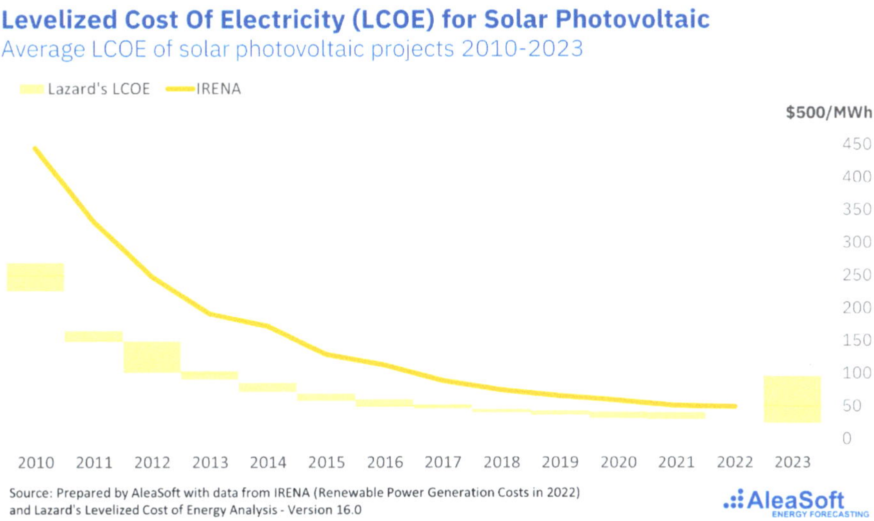

▲ Figure 48

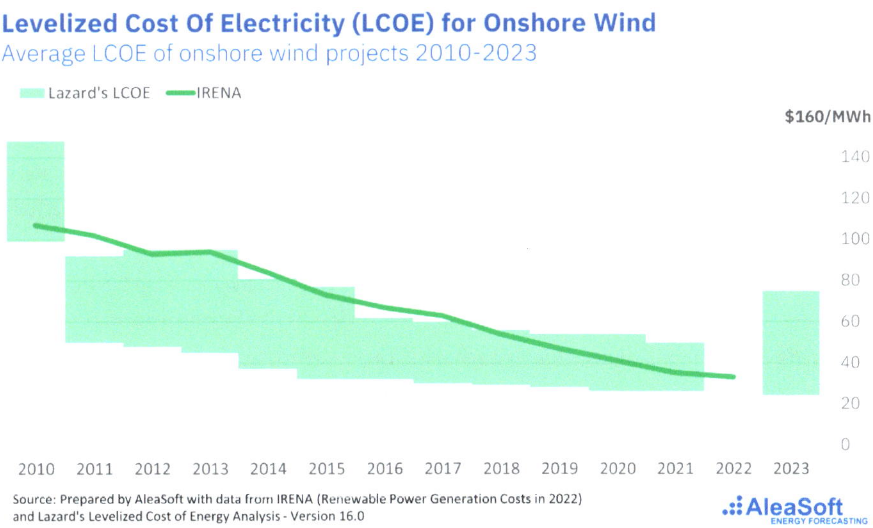

▲ Figure 49

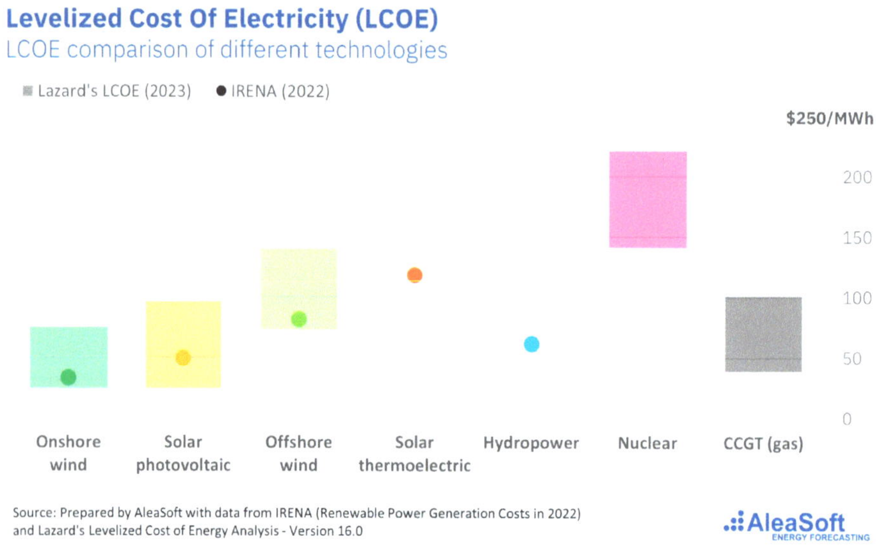

▲ Figure 50

India had the second most competitive LCOE for utility-scale solar PV and onshore wind projects in 2024. While the global average LCOE for onshore wind was US $0.034/kWh, India's LCOE for onshore wind in 2024 was around US $0.048/kWh or Rs. 3.8/kWh (based on an approximate conversion of 1 USD = 80 Rs).

India has emerged as a leader in renewable energy costs in the Asia-Pacific region, with solar costs falling to $38/MWh in 2024, according to the IRENA report. It is 14% cheaper than coal-fired power, according to Wood Mackenzie Power & Renewables. This cost competitiveness is partly due to India's large market size, high-quality solar resources, and increased competition in the renewable energy sector. This is expected to continue falling with further technological advancements and increased manufacturing capacity.

A dramatic 95% drop in the cost of solar photovoltaic modules, from over Rs. 200 per watt in 2010 to under Rs. 9 in 2024, is helping propel India towards a potential clean energy revolution, according to new research from the India Energy & Climate Centre (IECC) at the University of California, Berkeley.

The study *Plummeting Solar+Storage Auction Prices in India Unlock Affordable, Inflation-Proof 24/7 Clean Power*, suggested that the price plunge is not merely a technological milestone; it's a tipping point. It means India can now feasibly generate and store solar power for round-the-clock use at a price lower than most industrial electricity tariffs and new coal-fired power plants. Solar-plus-storage systems, the researchers found, can deliver 24/7 electricity with over 95% availability at under Rs. 6 per kilowatt hour.

Recent auctions in India have revealed record-breaking prices: Standalone battery storage was bid at as low as Rs. 2.8 lakh per megawatt (MW) per month (Rs. 0.39/kwh), and solar-plus-storage bids ranged from Rs. 3.1 to Rs. 3.5 per kWh, according to the research.

By optimising generation and storage, researchers estimated that a "flat block" of clean power—steady, round-the-clock electricity—can now be provided at Rs. 4–4.3 per kWh. Even with backup to ensure 100 per cent reliability, the cost remains under Rs. 4.7 per kWh—much lower than the average industrial electricity rate of Rs. 7.9 per kWh.

Figure 51 shows the decline in global PV module prices and the increase in global module production capacity, sourced from the International Energy Conservation Code (IECC). The IECC is widely adopted by states and municipalities across the United States and other countries to promote energy conservation and reduce energy consumption in buildings.

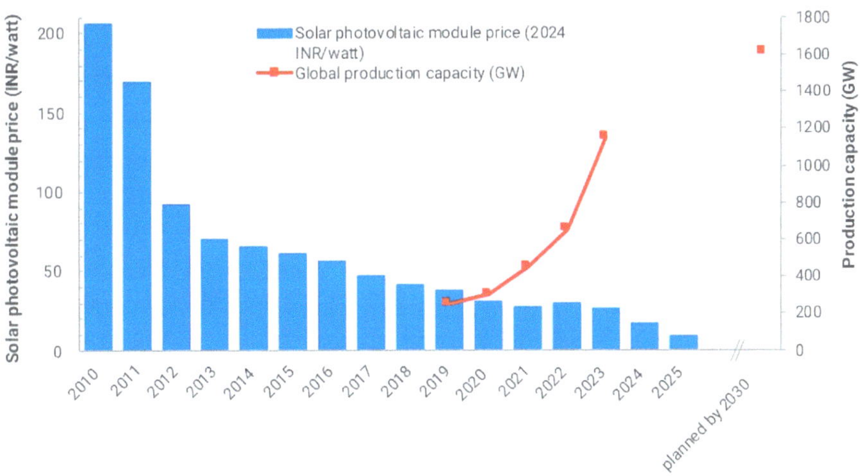

Global prices of solar photovoltaic modules have fallen to less than Rs 9 per watt (or $0.10 per watt). At the same time, global production capacity for solar PV modules has grown rapidly, reaching 1,150 GW by the end of 2023, with plans to increase it to 1,600 GW by 2030. IECC

▲ Figure 51

The shift isn't just about cheap solar panels. Battery storage technology has also advanced rapidly. Lithium-ion battery pack prices have fallen from Rs. 13,860 per kWh in 2020 to Rs. 8,388 per kWh in 2023, while their lifespan has increased to as many as 15,000 cycles without performance degradation. When solar and batteries are co-located on the same site, the resulting infrastructure savings make the economics even stronger.

India now has around 200 gigawatt-hours of battery manufacturing capacity under development, positioning it to become a global hub for energy storage, just as it did for low-cost solar deployment over the past decade, the paper said.

13.3 Breakeven Cost of Clean Hydrogen

The breakeven cost of green hydrogen fuel, meaning the price at which it matches the price of natural gas, is around $1.5/kg. For green hydrogen produced via electrolysis, costs currently range from $5 to $12 per kilogram, including delivery and fuelling, but could drop to $1.4 to $2.3 per kilogram by 2030 in some regions, potentially achieving cost parity with grey hydrogen by 2028 in the best areas. Conventional hydrogen from natural gas is much cheaper, at around $1.50 per kilogram, but it has a large carbon footprint.

Renewable hydrogen production costs continue to fall globally, more sharply than previously expected. Three factors are driving the cost reduction of clean hydrogen.

First, capex requirements for electrolysers are dropping. We expect a significant decline in electrolyser capex by 2030 to about USD 200–250/kW at the system level (including the electrolyser stack, voltage supply and rectifier, drying/purification, and compression to 30 bar), driven by accelerated cost roadmaps and faster scale-up of electrolyser supply chains. For example, several electrolyser manufacturers have announced near-term capacity scale-ups for a combined total of over approximately 3 GW per year globally.

Second, the LCOE is declining. Ongoing reductions in renewables cost to levels as much as 15% lower than previously expected, especially in regions with high solar irradiation (where renewables auctions continue to break record lows). The strongest reductions are expected in locations with optimal resources, including Spain, Chile, and the Middle East.

Third, utilisation levels continue to increase. Large-scale, integrated renewable hydrogen projects are achieving higher electrolyser utilisation levels. This performance is driven largely by the centralisation of production, a better mix of renewables (e.g., onshore wind and solar PV), and integrated design optimisation (e.g., oversizing renewables capacity versus electrolyser capacity for optimised utilisation).

Fourth factor, that is, process upgradation to reduce the cost of clean hydrogen, also helps in reaching the breakeven cost faster. The production of low-carbon H_2 also continues to gain momentum. Improvements include increased CO_2 capture rates for ATR from 95% in last year's report to 98%, coupled with potential capex reductions. Conducting ATR at higher temperatures can also increase methane-to-hydrogen conversion rates, resulting in a lower methane content in the product gas and further reducing emissions.

Introducing CO_2 costs can bring the earliest breakeven for clean hydrogen to 2028–2034, according to a McKinsey & Company report.

Figure 52 shows the production cost of hydrogen by different pathways. By 2030, the cost of green and blue hydrogen will be almost as much as that of grey hydrogen.

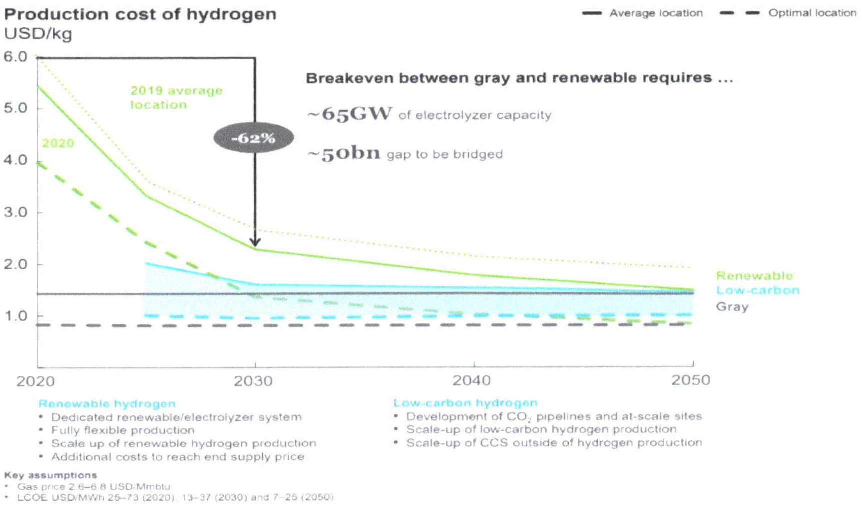

▲ Figure 52: Hydrogen Production Cost by different Pathways

The price of green hydrogen is expected to drop to around \$2/kg by 2030. RIL has an ambition to bring green hydrogen to \$1/kg by 2031–32. To compete with natural gas, the cost of green hydrogen should be around \$1/kg.

If we want to sell 1 MMBtu (Million British Thermal Units) of thermal energy (equivalent to 293.07 kWh) in hydrogen form, we need to sell 7.46 kg (293.07/39.3) of hydrogen. If the cost of natural gas is $10/MMBtu and we want to match the hydrogen cost to natural gas, the hydrogen cost should be $1.34/kg, based on its HHV. HHV of hydrogen is 39.3 kwh and of low heating value (LHV) 33.3 kwh per kg. Hope that India shall achieve this breakthrough by the mid-thirties.

Electrolysers do not operate in laboratory conditions in the field. They operate under highly unstable renewable energy profiles in a torturous climatic condition of different geography. Solar output is a bell curve cyclic order. Wind output is rapid and highly unpredictive. The grid gives pricing signals. The composition of these three elements determines capacity utilization of an electrolyser, and capacity utilization determines hydrogen cost more than marginal efficiency gains.

Hydrogen economics is therefore not only electrochemistry but systems engineering as well. If the load curve is unstable, the material degradation accelerates. If the ramping profile is aggressive, the life of the electrolyser reduces fast. If electricity pricing is poorly structured, Levelized Cost of Hydrogen will rise exponentially. Very little improvement (say 1 or 2%) in stack energy efficiency, cannot compensate for a plant operating at as low as capacity of 35% utilization. In other words, electrochemistry sets the ceiling, and the load curves determine the floor.

Hydrogen economics is therefore more about systems engineering consisting of Renewable intermittency, PPA (Power Purchase Agreement) structure, Curtailment management, Storage buffering and Grid Tariff design.

14 Hydrogen Economy

14.1 Concept of Hydrogen Economy

The concept of a society that uses hydrogen as its primary means of energy storage was first proposed by geneticist JBS Haldane in 1923. Anticipating the exhaustion of Britain›s coal reserves for power generation, Haldane proposed a network of wind turbines to produce hydrogen and oxygen for long-term energy storage through electrolysis, to help address renewable power's variable output.

The term "hydrogen economy" was coined by John Bockris during a talk he gave in 1970 at General Motors (GM) Technical Centre. Bockris viewed it as an economy in which hydrogen, underpinned by nuclear and solar power, would help address growing concern about fossil fuel depletion and environmental pollution by serving as an energy carrier for end-uses in which electrification was not suitable.

Modern interest in the hydrogen economy can generally be traced to a 1970 technical report by Lawrence W. Jones of the University of Michigan, in which he echoed Bockris› dual rationale of addressing energy security and environmental challenges. Unlike Haldane and Bockris, Jones focused solely on nuclear power as the energy source for electrolysis and, principally, on the use of hydrogen in transport, where he regarded aviation and heavy goods transport as the top priorities.

In 1974, the International Energy Agency (IEA) and the International Association for Hydrogen Energy (IAHE) were established. These organisations had several initiatives, including advocating for national strategies to increase interest and visibility of the hydrogen economy.

The hydrogen economy refers to a future energy system where hydrogen, produced using sustainable methods, serves as a major energy carrier, potentially replacing fossil fuels and reducing greenhouse gas emissions. It involves the production, storage, distribution, and utilisation of hydrogen across various sectors like transportation, industry, and power generation.

A hydrogen economy is a vision in which hydrogen energy is marketed and utilised like any other energy carrier, adding value to the nation.

Infrastructure development and strategic planning shall be required. IEA has observed that the concept of the hydrogen economy can be translated into reality when green hydrogen dominates the fuel market in the transport sector.

Vision of a Hydrogen Economy

- The hydrogen economy envisions a shift from fossil fuels to hydrogen as a clean energy carrier.
- Hydrogen Production: Hydrogen can be produced from various sources, including renewable energy through electrolysis of water, natural gas reformation (with carbon capture), and even biomass. In gas reformation, a carbon capture facility has to be added to produce clean hydrogen. Nuclear energy can supply clean power and heat to effectively reduce the cost of clean hydrogen.
- Hydrogen Storage and Distribution: Hydrogen can be stored in various forms (gaseous, liquid, or as a chemical hydride) and transported via pipelines, trucks, or ships.
- Hydrogen Utilisation: Hydrogen can be used in fuel cells to generate electricity, in combustion engines of the transport sector, or as a feedstock for various industrial processes.
- Sectors: The hydrogen economy aims to decarbonise transportation through internal combustion or fuel cell vehicles, heavy industry through the application of green hydrogen in steel and cement manufacture, and power generation through seasonal energy storage, which reduces carbon emissions.

14.2 Benefits of a Hydrogen Economy:

Reduced Greenhouse Gas Emissions

Hydrogen is a clean energy carrier, especially when produced from renewable sources, significantly reducing carbon emissions. It emits no CO_2 on combustion or conversion to electricity.

Energy Diversification

It can diversify energy sources, reducing reliance on fossil fuels and enhancing energy security.

Hydrogen can be used to store excess renewable energy and provide a flexible energy source for various applications.

Hydrogen can play a crucial role in decarbonising sectors where direct electrification is difficult or expensive, like heavy industry. It can be delivered through pipelines and packaging, for distribution to any part of the geography, similar to natural gas and LPG.

14.3 Challenges of the Hydrogen Economy

The current challenge for the global energy economy is to implement a large-scale, low-emission alternative to fossil fuels that is economically viable, safe, and sustainable. The key to progress in reducing pollutant emissions and combating energy poverty is the energy source and its price. Balancing greenhouse gas emissions and their removals will enable the global economy to transition to zero-emission activities, but it will require a comprehensive approach and a long-term strategy based on innovative technologies.

Investments in the hydrogen economy will enable the use of hydrogen as an energy carrier to help meet the growing energy demand due to the high pace of development of civilisation and depleting fossil fuel resources. Due to the possibility of ecologically obtaining energy from renewable energy sources to produce green hydrogen, many research and development efforts are underway to identify new electrode materials capable of catalytic participation in reducing the energy barrier for obtaining green hydrogen through the emission-free process of electrolysis. The current technologies used for green hydrogen production primarily involve 4 types of electrolysers: alkaline electrolysers (AEs), proton exchange membrane (PEM) electrolysers, solid oxide electrolysers (SOEs), and anion exchange membrane (AEM) electrolysers.

In addition to electrolyser technologies, there are other innovative methods being researched and developed for green hydrogen production, including photoelectrochemical (PEC) water splitting, which uses solar energy directly to drive electrolysis; biological hydrogen production, which involves using microorganisms to produce hydrogen through processes like dark fermentation and biophotolysis; and high-temperature electrolysis (HTE) through SOE, that can use heat from nuclear reactors or concentrated solar power to improve efficiency. The choice of technology for green hydrogen production depends on various factors, including the availability and cost of renewable energy, production scale, system efficiency, capital and operating costs, and the specific application or end use of the hydrogen. As demand for green hydrogen grows, these technologies are expected to continue evolving and become more cost-effective and efficient.

Tasks and Challenges of the Hydrogen Economy Towards Green Hydrogen

Figure 53 presents the main tasks of the hydrogen economy. R&D conducted within the functional stages of the hydrogen economy concerns:

a. Green hydrogen production (stage I)

b. Green hydrogen purification and transmission (stage II)

c. Green hydrogen storage (stage III) and

d. Conversion of chemical energy of green hydrogen into electricity using fuel cells (stage IV)

Stage I primarily involves developing low-emission or emission-free methods for hydrogen production. Their costs must be significantly reduced for widespread use and mass production for hydrogen energy to be competitive with conventional fossil-fuel-based energy. Hydrogen can be transferred via gas pipelines, as with natural gas (stage II), or transported under pressure in tankers (stage III), and finally, it can be converted to electricity via a fuel cell (stage IV).

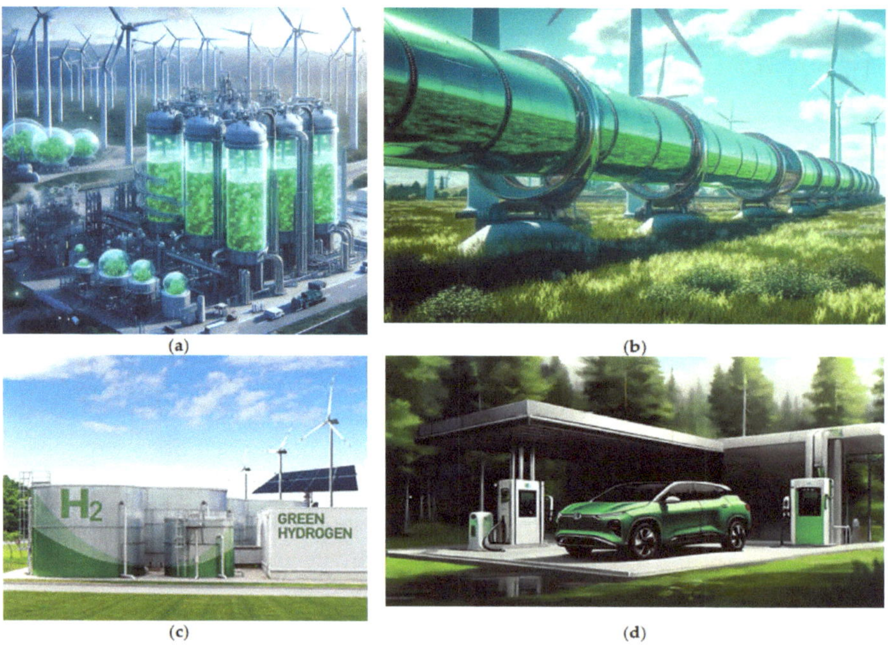

Figure 3. Hydrogen Economy (HE) tasks towards green hydrogen: (**a**) A concept of green hydrogen production through water electrolysis (stage I) [81]; (**b**) Green hydrogen purification and transmission (stage II) [82]; (**c**) Green hydrogen storage (stage III) [83]; (**d**) Green hydrogen conversion by fuel cells (stage IV) [84].

▲ Figure 53: Hydrogen economy tasks for green hydrogen

Hydrogen storage is the most difficult technical barrier to overcome, which hinders the implementation of hydrogen technologies at an industrial scale. Presently, expensive pressure vessels and less durable cryogenic tanks are used to store hydrogen. In stage III, it is necessary to develop methods for storing hydrogen, especially for transport applications. Progress in technologies for obtaining materials for hydrogen storage is a factor determining the economic success of using hydrogen to power cars. Therefore, research is being conducted on the safe storage of hydrogen in the crystalline structures of metals and their alloys, called metal hydrides. The most important indicator for hydrogen use in fuel cells in stage IV is purity, which must meet the 5.0 standard (99.999%). Even the smallest traces of hydrogen contamination may damage the efficiency and durability of the fuel cell system.

The main bottlenecks why green hydrogen is not being able to penetrate the global energy market are:

High Production Costs

Currently, the cost of producing clean hydrogen (from renewable sources) is high. Hydrogen production costs for different pathways in 2024 and projected for 2030 under the IEA-defined policy scenario are shown in Figure 54. The stated policies scenario takes into account the existing policies and strategic planning of states, not the stringent policies required to meet the climate target as per the UNFCCC in the Paris Convention. The cost ranges reflect regional differences in fuel prices, renewable energy costs, CO_2 prices, technology, capex, opex, and the cost of capital. This figure has been taken from IEA's Global Hydrogen Review 2025.

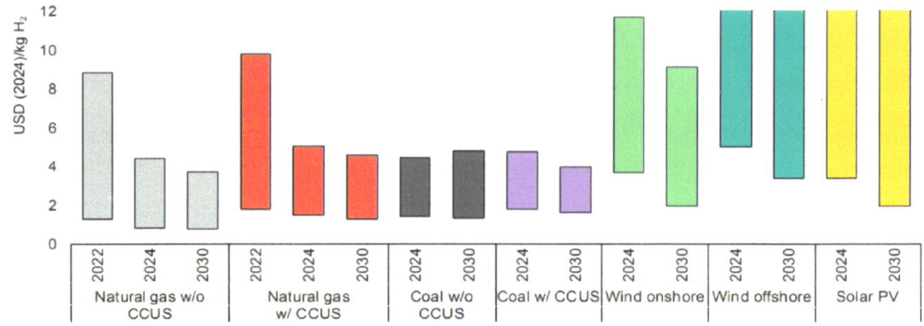

Notes: CCUS = carbon capture, utilisation and storage; w/ = with; w/o = without. Cost ranges reflect regional differences in fossil fuel prices, renewable costs, CO_2 prices, technology CAPEX and OPEX as well as cost of capital. Natural gas price is USD 5.7-50/MBtu for 2022, USD 1.7-25/MBtu for 2024 and USD 1.6-21/MBtu in the Stated Policies Scenario (STEPS) in 2030. Coal price is USD 21-260/t for 2024 and USD 22-200/t for the STEPS in 2030. CO_2 price is USD 0-90/t CO_2 for 2022, USD 0-70/t CO_2 for 2024 and USD 0-106/t CO_2 for the STEPS in 2030. The levelised production cost of solar PV electricity is USD 20-110/MWh for 2024 and USD 13-85/MWh for the STEPS in 2030, with a capacity factor of 12-35%. The levelised production cost of onshore wind electricity is USD 25-100/MWh for 2024, USD 23-90/MWh for the STEPS in 2030, with a capacity factor of 30-53%. The offshore wind electricity levelised production cost is USD 52-270/MWh for 2024, USD 40-165/MWh for the STEPS in 2030, with a capacity factor of 30-67%. For China, electrolyser CAPEX is USD 900/kW in 2024 and USD 675/kW for the STEPS in 2030, whereas for the rest of the world, it is USD 2 300/kW in 2024 and USD 1 600/kW for the STEPS in 2030. Electrolyser CAPEX includes the electrolyser system, balance of plant, engineering, procurement and construction (EPC) and contingencies; electrolyser capacity factor is assumed to be the same as the renewable power plant. The cost of capital is 6-20%. Figure capped at USD 12/kg H_2, although some production routes reach higher values. Water cost is not included. More details on the cost of producing low-emissions hydrogen using different technologies and in different regions can be found in the new Hydrogen Tracker available on the IEA website.

Sources: IEA analysis based on data collected through a survey to original equipment manufacturers, EPC companies and project developers, data from McKinsey & Company and the Hydrogen Council; BloombergNEF, Argus Media Group, All rights reserved, NETL (2022); IEA GHG (2017).

The recent drop in fossil fuel prices has increased the cost gap between low-emissions hydrogen production and unabated fossil-based routes.

▲ Figure 54: Hydrogen production cost by pathway, 2024, and in the Stated Policies Scenario 2030

It is expected that by 2030, the production cost of green hydrogen from onshore wind and solar PV would be \$2/kg, which is about \$15 per MMBtu. Natural gas prices in Europe (TTF benchmark) and Asia (JKM benchmark) have been around \$9– \$10 per MMBtu in August-September 2025 due to supply and demand factors. With this price of natural gas, H_2 cost has to come down \$1 to 1.5 per kg to compete with natural gas in the energy market.

Infrastructure Development

Building a comprehensive hydrogen infrastructure (production, storage, distribution, and refuelling stations) requires significant investment and time.

Public Acceptance

Public awareness and acceptance of hydrogen as a safe and reliable energy source are crucial for its widespread adoption. Risk mitigation techniques and affordability must be developed to earn public acceptance.

Technological Maturity

Further advancements in hydrogen production, storage, and utilisation technologies are needed to improve efficiency and reduce costs. These technologies are currently in the development stage and have yet to mature.

The hydrogen economy is a long-term vision, and its success depends on overcoming these challenges and realising its potential to contribute to a sustainable and low-carbon energy future. Hydrogen demand is currently concentrated in refining, ammonia sectors, and industrial feedstocks, but there's growing interest in using it for transportation, energy storage, and other sectors that need decarbonisation.

PWC and other experts predict limited but steady growth in hydrogen demand until 2030 for several reasons.

First, current hydrogen projects under construction and in operation, despite growing capacities, are at a pre-commercial phase and have limited electrolyser capacities, typically well below 50MW. Proposed plants have electrolyser capacities of 100MW or more, but these are still small compared with current grey hydrogen production plants.

Second, building the infrastructure for large-scale hydrogen use, such as pipelines or export/import terminals, will take many years—it takes 7 to 12 years to plan and build a pipeline, for example. Ideally, the required infrastructure will be built in parallel with growing hydrogen demand and falling costs, ensuring that by 2030, hydrogen can be traded and transported in the necessary quantities. All medium- and high-ambition scenarios see stronger hydrogen demand from 2030, with another strong increase from 2035 onwards.

McKinsey & Company has extensively analysed the potential of the hydrogen economy, particularly in the context of the global energy transition and decarbonisation efforts. They highlight the potential of hydrogen, especially clean hydrogen produced from renewable sources, to play a significant role in decarbonising industries like aviation, shipping, steel, and chemicals. McKinsey's analysis suggests that clean hydrogen could account for a substantial portion of total hydrogen demand by 2050. According to him, demand is projected to grow 4 to six-fold by 2050 from today's 96 tonnes/day. Hydrogen has the potential to cut annual global emissions by up to 20 percent by 2050.

Bloomberg NEF predicted that over half of the total hydrogen supply in 2030 will come from electrolysis. Blue H_2 plays a significant role. Around 95 gigawatts (GW) of electrolysers could be operational globally by the end of 2030.

The estimates for global hydrogen production in 2050 vary significantly, ranging from under 200 Mt (million tonnes) to over 600 Mt, with 85% projected to come from a low-carbon pathway. These projections are highly dependent on underlying assumptions about the pace of the energy transition, technological advancements, infrastructure development, and policy frameworks.

The global hydrogen economy is a developing field focused on using hydrogen as a clean energy carrier. It's gaining traction for its potential to decarbonise sectors such as transportation, industry, and power generation, while aligning with global climate goals. While currently dominated by fossil-fuel-based hydrogen production, there's a growing emphasis on low-carbon emission hydrogen production and utilisation.

Takeaways from future hydrogen production estimates, made by major global players, are:

- Significant growth is expected: All major players agree that hydrogen production will increase dramatically by 2050 to the level of 600 tpd.
- Clean hydrogen will dominate: The majority of this increased production is expected to come from low-carbon sources, primarily green hydrogen (produced using renewable energy) and blue hydrogen (produced from natural gas with carbon capture). 85% is estimated to be clean hydrogen with a major contribution from green hydrogen.
- Industry will be a major driver: Existing industrial applications, such as refining and chemical production, mostly ammonia, are expected to be early adopters of clean hydrogen, transitioning from grey hydrogen. Heavy industries such as steel and cement will drive demand from 2040 onwards. The transport sector will offtake significantly.

Globally, the transport sector accounts for around a quarter to a third of total final energy consumption. For some industrialised countries, this figure is closer to 30%. In 2021, road transport accounted for nearly 78% of energy consumption in all transport sectors, followed by marine transport (11%), aviation (8%), and rail (3%).

- New applications will emerge: By 2050, hydrogen is expected to play an increasingly important role in sectors like steel production, transportation (especially heavy-duty and long-haul), and potentially even heating and power generation.
- Challenges remain: The cost of clean hydrogen production, the pace of infrastructure development, and competition from other decarbonisation technologies are significant hurdles to realising the full potential of hydrogen.

- Regional disparities: Different regions are expected to adopt hydrogen at different rates, with factors like natural resource availability, energy policies, and existing infrastructure influencing the transition.

14.4 Key Recommendation of the International Energy Agency to Scale Up Hydrogen Production.

At the request of the government of Japan under its G20 presidency, the IEA produced a landmark report in 2019 on the current state of hydrogen production and offered guidance on its future development. The report finds that clean hydrogen is currently enjoying unprecedented political and business momentum, with the number of policies and projects around the world expanding rapidly. It concludes that now is the time to scale up technologies and reduce costs to enable hydrogen to become widely used. The pragmatic, actionable recommendations provided by the IEA to governments and industry will help them take full advantage of this increasing momentum. These are 7 recommendations from the IEA to scale up hydrogen production, as outlined in the report *The Future of Hydrogen – Seizing Today's Opportunity*, published in June 2019. These are:

1. Establish a role for hydrogen in long-term energy strategies. National, regional, and city governments can guide future expectations. Companies should also have clear long-term goals in key sectors, which include refining, chemicals, iron and steel, freight and long-distance transport, buildings, and power generation and storage.
2. Stimulate commercial demand for clean hydrogen. Clean hydrogen technologies are available, but costs remain challenging. Policies that create sustainable markets for clean hydrogen, especially to reduce emissions from fossil fuel-based hydrogen, are needed to underpin investments by suppliers, distributors, and users. By scaling up supply chains, these investments can drive cost reductions, whether from low-carbon electricity or fossil fuels with CCUS.
3. Address investment risks of first-movers. New applications for hydrogen, as well as clean hydrogen supply and infrastructure projects, stand at the riskiest point of the deployment curve. Targeted and time-limited loans, guarantees, and other tools can help the private sector to invest, learn, and share risks and rewards.
4. Support R&D to bring down costs. Alongside cost reductions from economies of scale, R&D is crucial for lowering costs and improving performance, including for fuel cells, hydrogen-based fuels, and electrolysers.

5. **5.** Eliminate unnecessary regulatory barriers and harmonise standards. Project developers face hurdles where regulations and permitting requirements are in place. Harmonising standards, including for equipment, safety, and emissions certification from different sources, has yet to be established and streamlined. Hydrogen's complex supply chains mean governments, companies, communities, and civil society need to consult regularly.
6. Engage internationally and track progress. Enhanced international cooperation is needed across the board, especially in standards, the sharing of good practices, and cross-border infrastructure. Hydrogen production and use need to be monitored and reported regularly to track progress towards long-term goals.
7. Focus on 4 key opportunities to further increase momentum over the next decade. By building on current policies, infrastructure, and skills, these mutually supportive opportunities can help to scale up infrastructure development, enhance investor confidence, and lower costs:
    - Make the most of existing industrial ports to turn them into hubs for lower-cost, lower-carbon hydrogen.
    - Use existing gas infrastructure to spur new clean hydrogen supplies.
    - Support transport fleets, freight, and corridors to make fuel cell vehicles more competitive.
    - Establish the first shipping routes to kick-start the international hydrogen trade.

15 Hydrogen Is Used in Many Other Ways

Hydrogen and electricity can combine to create a powerful hybrid energy system, making our energy supply greener, smoother, and cheaper. Electrons (electricity) can be converted into molecules (hydrogen), which can be transported more conveniently, at lower cost, and with fewer losses.

1. Hydrogen is an energy storage like a battery. To tap the full potential of solar and wind, the energy produced can be stored as hydrogen, which can be used in a fuel cell for electricity, as fuel for internal combustion, or as a feedstock for chemicals. The gap arising from the solar and wind power output variations and power demand can be bridged with this hydrogen storage. This will help harness the full potential of solar and wind. Hydrogen storage can cost less than having battery storage, pumped hydro, and many expensive energy storage systems. The output can be stratified from intermittent supply through solar and wind, and the optimal potential can be drawn. A combination of an electrolyser and fuel cell has been reported to be better than battery storage when the storage/consumption cycle is very low, say once a week.
2. Solar and wind power can split water at their generation point itself to produce hydrogen, which then can be transported through a pipeline or any other suitable means.

   The Oyster project has planned to bring together wind power companies Orsted and Siemens Gamesa, electrolyser maker ITM Power, and the consultancy Element Energy. They will develop and test an MW-scale compact electrolyser, hardened to withstand the marine environment, integrated with a single turbine and a desalination system to electrolyse seawater.

   These 3 projects are a cooperation between molecules and electrons. This power couple is coming together to transport and store renewable electricity from one of the cheapest renewable sources in Europe, using existing pipeline infrastructure where possible, limiting congestion on the power grid, and providing customers who need hydrogen with cheap, decarbonised sort.

3. One of the challenges with electricity is long-distance transport, which is expensive and wastes energy.

   Hydrogen pipelines are cheaper than electricity transmission lines. Hydrogen pipelines can effectively transport renewable hydrogen across long distances. They can transport 10 times the energy at one-eighth the cost associated with electricity transmission lines. Furthermore, hydrogen pipelines have a longer lifespan than electricity transmission lines and offer dual functionality, serving as a transmission and storage medium for green energy.

   It's true that we lose 30% of renewable power when we generate green hydrogen, as electrolysis is about 70% energy-efficient, but turning renewables into hydrogen can still be cost-effective, especially in sectors that are technically challenging or expensive to electrify. Hydrogen by pipeline will also be appealing where it is difficult to build power lines.

   The Indian western offshore oil and gas field has a large gas pipeline network. Installing wind turbines on abandoned platforms to harness wind power and convert it to hydrogen locally via an electrolyser is worth considering. Hydrogen can be blended in gas network at a suitable point. The suitability of the structure of the abandoned well head platform needs to be examined. No separate pipeline for hydrogen transport shall be needed; it can be blended up to 20% in existing natural gas pipelines safely.
4. Swapping of energy, where solar/wind is more productive, and gas is more expensive. The PPWS (Put Panels Where It's Sunny) concept involves a kind of virtual export of renewable energy. Put panels where solar is more productive in terms of energy output and cost, and replace natural gas energy there with renewables so that the gas-equivalent energy can be inter-relocated.

16 Bringing It All Together

The combustion of fossils, which accounts for more than 82% of global primary energy, emits a huge amount of carbon dioxide. It is estimated that around 40 billion tonnes of carbon dioxide (CO_2) is emitted per year from industries globally. Other gases with high GWP, such as methane, nitrous oxide, and carbon hexafluoride, are also emitted alongside CO_2 during the combustion of fossil fuels. If we take other gases into account with their GWP, the total greenhouse gas emission comes out to be around 51 billion tonnes per annum of CO_2-equivalent.

CO_2 is the primary cause of global warming. Rapid industrialisation, particularly since the 1970s onwards, has led to an exponential rise in CO_2 emissions from the extensive combustion of coal, oil, and gas to supply energy. In 1950, the world emitted 6 billion tonnes of CO_2. By 1990, this had almost quadrupled, reaching more than 20 billion tonnes. Emissions have continued to grow rapidly, and by 2024, 40 billion tonnes of CO_2 was emitted annually, a rise of more than 6 times from the 1950s level.

It has been well established through 40 years of scientific study from the 1950s to the 1990s that rising atmospheric CO_2 concentration leads to an increase in global temperature, which in turn causes global warming.

Global warming is a threat to human survival because it disrupts Earth's ecological balance, leading to extreme weather, melting glaciers, sea-level rise, floods, droughts, and other adverse consequences. Global efforts were made to reduce emissions of greenhouse gases, including those other than carbon dioxide, like methane, nitrous oxide, and sulphur hexafluoride, through the Kyoto Protocol held in Japan in 1997. But it could not bring any measurable results because of the non-participation of major nations like the US, China, and Australia. Subsequently, the Paris Convention on Climate Change was held in 2015, with the participation of 195 nations all over the world. The Paris Convention unanimously agreed to reduce and reverse global temperature rise and committed to limiting it to 2°C by the end of the century, with further efforts to limit it to 1.5°C. NDC (Nationally Determined Contribution), a key mechanism for monitoring progress, was developed to limit fossil fuel combustion and track CO_2 emissions. Renewable energy sources, as alternatives to fossil fuels, were encouraged and incentivised. Renewables and low-carbon energy sources like solar, wind, biomass, and nuclear were introduced into the energy mix, which is developing at a very fast pace. Growth has been exponential over the last 5 years. India is leading in the race for renewable energy installation and utilisation, aiming to reduce fossil fuel consumption by cutting down fossil-based installed power generation capacity by 50%, replacing it with non-fossil, and reducing CO_2 emissions by one billion tonnes per annum by the year 2030.

Two main renewable sources, solar and wind, are developing at a very fast pace. But these are intermittent, depending upon weather, location, and time (day/night). Solar is more active during the day, and wind is more active at night. Therefore, these 2 can be supplementary to each other to a large extent. To tap their full potential, an energy storage system, such as a battery, a water pump, or an energy carrier like hydrogen, will be required.

Solar and wind together constitute 91% of the total renewable potential in India. Total renewable energy generation potential in India, including all sources, is reported to be 2,110 GW, consisting of 749 GW (36%) solar, 1,164 GW (55%) wind (both onshore and offshore at 100-metre hub height), 133 GW (6%) Large Hydro, and balance by biomass (30 GW) and geothermal (10GW). Solar and wind energy can supplement total energy needs, but they are not enough to replace fossil fuels fully.

Hydrogen was found to be the most suitable alternative to fossil fuels, which has the potential to replace them. This is because of 3 vital reasons that other fuels don't have. Firstly, it does not emit CO_2 on combustion, but only water vapour, and hence is a completely clean fuel; secondly, it is available in abundance on Earth in the form of water; and thirdly, it can be made accessible to all parts of the geography like petroleum products. Hydrogen is not a primary energy source like fossil fuels, but rather an energy carrier like diesel, gasoline, and other transport fuels, because hydrogen is derived from a primary energy source.

Hydrogen can be stored, transported, distributed, and dispensed anywhere, in the same way as oil, gas, or coal. Transportation of H_2 through a pipeline is much more economical than electric transmission. It can carry 10 times as much energy at one-eighth the cost of electric transmission. The lifespan and maintenance costs of a hydrogen pipeline are much lower than those of electric transmission lines.

Hydrogen is also available in organic materials, called biomass. Hydrogen can be produced from biomass through thermogenic (gasification) or biogenic processes. The combustion of biomass produces CO_2, but this is reabsorbed by plants through the photosynthesis process, and there is no net accumulation in the atmosphere, hence leaving no carbon footprint. Therefore, biomass combustion is a carbon neutral.

It is worth mentioning here that, as per BBC Science Focus Magazine's publication, a hydrogen reserve has been found underground. Geologic hydrogen is formed during natural geochemical reformation processes in the Earth and, so far, has been discovered in just a few places, including Albania and Mali. A new study, published in the journal *Science Advances*, estimates that these reserves are stretched all around the globe. The Alps mountain range and the Pyrenees mountain range of mountains in Europe are reported to contain a hydrogen reserve underground. But no substantial evidence of hydrogen exploration and production on a large scale is known.

In pursuance of our ambition to replace fossil fuels with hydrogen, hydrogen production through water electrolysis is the best option. Biomass availability, as a feedstock for hydrogen production, is much less than what is needed to replace even a small fraction of fossil consumption.

We need a renewable source of energy to feed the electrolyser for producing green hydrogen. Fossil emits about 9 kg of CO_2 to produce just 1 kg of hydrogen, which does not serve the purpose of decarbonising the planet. A non-emitting CO_2 source of energy, such as renewables or nuclear, will be needed for this purpose.

Blue hydrogen, a form of clean hydrogen, is derived from fossil fuels by capturing carbon dioxide, utilising and storing it, called CCUS technology. The cost of adapting carbon capture technology in a power plant varies widely, but generally adds significant expense to electricity generation. Adding a carbon capture facility to a natural gas power plant can increase the cost per megawatt-hour from about \$55 to \$85, including project capital, energy penalty, and operating costs. It has been reported that about 162 kwh is consumed per ton of CO_2 capture. The cost of per metric tonne of CO_2 captured also varies, ranging from \$40- \$120 per tonne from power plants, as per the IEA report in 2021. The large variation is on account of changes in the source and concentration of CO_2 in emission.

A nuclear power plant can be a big source of clean energy. It is argued that because a nuclear plant is a potent source of radiation, the consequence of meltdown, if it happens, is highly fatal and has a long-term impact on human society. Hazard of Chornobyl, Three Mile Island, and Fukushima are referred to. But the statistics say that fatalities due to toxic gas emissions are more than those due to nuclear radiation in the long run. Three-fourths of the power generation in France comes from nuclear power plants. The US alone runs

100 GW of nuclear plants. India has installed only an 8.8 MW nuclear plant till now, but has a plan to go to 100 mw by the middle of the century.

Splitting of water molecules at high steam temperature, generated from a nuclear power plant, shall be more efficient, even more than 90%, with the application of a solid oxide electrolyser. But due to the scarcity of nuclear fuel (U-235), high capital expenditure, and the risks involved in handling it, nuclear power plants are not preferred in India to a large extent, whereas the US, China, France, Russia, and South Korea use them extensively. As per the recent nuclear policy of India, it is planning to go for nuclear energy in a broader way.

India does not have a uranium reserve, fuel for a nuclear plant. The 48-nation NSG (Nuclear Suppliers Group) granted India a waiver on 6 September 2008, allowing it to access civilian nuclear technology and fuel from other countries. Subsequently, as of 2016, India has signed civil nuclear agreements with 14 countries, including Russia, Japan, France, South Korea, and the US, to trade nuclear fuel. India has been making advances in the field of thorium-based fuels by developing a prototype breeder atomic reactor that uses thorium and low-enrichment uranium. This is a key part of India's three-stage nuclear power programme, conceived by H.J. Bhabha. Thorium is abundantly available in the form of monazite in the coastal sands of Kerala and Orissa.

1 kilogram of uranium can produce as much energy as 3 million kilograms of coal. Nuclear energy remains one of the most energy-dense and clean power sources available today. With zero carbon emissions during operation, nuclear power is gaining renewed attention as nations seek sustainable alternatives to fossil fuels.

Modern nuclear reactors, including Small Modular Reactors (SMRs), are safer and more efficient, reducing waste and offering scalability. While concerns about safety and disposal persist. Advancements in technology and safety regulation in design, construction, and operation have significantly improved risk mitigation.

As the world pushes towards net-zero targets, nuclear energy is being reconsidered as a part of the global clean energy mix, more broadly.

India is aggressively pursuing an expansion of its nuclear power capacity, aiming to reach 100 GW by 2047 through legislative and policy changes. This ambitious goal is part of a broader strategy to transition to a green energy future and reduce reliance on fossil fuels. Key aspects of this plan include legislative changes, international partnerships, and the development of both indigenous and foreign technologies. Though the capital expenditure for a nuclear plant is high (as much as $1,700 per kW, according to the World Nuclear Association), the operating cost is very low.

There are other sources of clean energy generation that are emerging but are still in their infancy and may take decades to come to realities. These are all at an early stage in the experimental process and are neither technically nor economically feasible for public utilities at this time. These innovative technologies, often discussed in the search for clean energy, include fusion technology, solar satellites, and salt reactors. These sources of energy, as and when they come up, shall be the best feed to green hydrogen production with the least carbon emission.

1. Fusion is the process of smashing nucleus of deuterium and tritium to a single nucleus of helium, which is initiated at a very high temperature, 10 times that of the core of Earth. The energy released during fusion is in the form of plasma, which is captured through a powerful magnetic or inertial confinement.
2. In a solar satellite, solar panels are installed in space to capture full solar radiation. Only 30% of solar radiation reaches Earth; the rest is absorbed by atmospheric atoms and molecules. From space, solar energy is transmitted by laser or microwave to an Earth power station through a tran*sd*ucer, where it can be fed to the grid for transmission and distribution.
3. In a salt reactor, radioactive salts (uranium, thorium, or plutonium fluoride salts dissolved in a mixture of lithium and beryllium fluorides) are mixed in the reactor, which generates heat energy, which is then used to generate steam at high temperature and pressure to rotate the turbine. To what extent this will be a source of future energy and to what extent it can meet global demand are currently uncertain. This technology was tried earlier, but could not proceed due to corrosion issues. A modified Natrium reactor is being tested by TerraPower.

Hydrogen is the most viable option for replacing fossil fuels to a large extent. To do this, hydrogen has to enter the global energy market and become part of the global energy mix for public utilities. What is constraining us from bringing hydrogen to the fuel market is the cost and risks involved in handling it.

Awareness of global warming and commitment to mitigating it have been exceptionally effective in promoting clean hydrogen as an alternative to fossil fuels globally. The exponential rise in solar and wind power installations and production has resulted in a sharp decline in renewable energy costs. Improvements in electrolyser technology and costs, and government initiatives and encouragement are sharply bringing down the levelised cost of clean hydrogen. R&D works in science and engineering to develop new materials and technologies in photovoltaic cells and electrolysers, new ideas emerging in finding ways and means to reduce risks in handling of hydrogen, are bringing highly encouraging results

It has been predicted that by mid-thirties, the levelised cost of green hydrogen will be around $1.5/kg, comparable to grey hydrogen. RIL has the ambition to bring down green hydrogen cost to $1/kg by 2032.

It has been estimated that demand for green hydrogen will be initially driven by industries, mainly refineries and the ammonia industry, until 2040, and later by energy demand, mostly in the mobility sector. The DRI (sponge iron) and cement industries are potential consumers of clean hydrogen in the global decarbonisation campaign. Various agencies have predicted different figures for hydrogen demand by 2050, but most estimates range from 520 to 600 million tonnes per annum. IEA has predicted 14% share of hydrogen in the 2050 energy mix.

Critical metals will play a very important role in the growth of hydrogen and may be decisive. Solar and wind turbines, batteries, electrolysers, and fuel cells, which are the ways and means of producing green hydrogen and converting hydrogen to electricity, are heavily dependent upon critical metals, including rare-earth magnets such as zirconium, iridium, yttrium, palladium, graphite, lithium, nickel, and many others. The availability of such elements is very limited and centred in certain regions such as China, Brazil, Chile, Argentina, Congo, and South Africa. India has to develop a strategic approach, indigenous exploration and production, and R&D to meet its demand.

The world today faces twin challenges: the growing demand for energy and the urgent need to decarbonise the planet. In this context, hydrogen is a beacon of hope and prosperity. With demand for energy growing and India's net-zero target set for 2070, hydrogen is being positioned not only as a rocket fuel by NASA, but also as a driver of transport, power generation, and other industrial decarbonisation processes. Hydrogen IC engines and hydrogen carriers, like methanol and ammonia, are particularly vital for heavy-duty transport and marine sectors, where batteries fall short.

India must accelerate its journey towards a hydrogen economy in pursuance of achieving zero carbon emissions. Methanol and ammonia can serve as effective hydrogen carriers to overcome transport and storage challenges. High cost of renewable energy, electrolyser cost, hydrogen handling, and fuel cell create a significant gap between hydrogen production and end use.

Electrodes, electrolytes, and catalysts are critical as well as major cost-bearing elements responsible for energy efficiency and voltage efficiency. Most R&D efforts focus on developing an efficient and cost-effective alternative to electrodes, electrolytes, and catalysts. Technological advancements in water electrolysis focus on boosting efficiency, lowering costs, and improving integration with renewables. Better catalysts, advanced membranes, and optimised stack designs are allowing the use of seawater and unlocking scalable green hydrogen production for decarbonisation. Key progress includes high-temperature SOECs, semi-vapour electrolysis for seawater, and better integration with variable renewable sources.

Efforts in the field of electrolysis of water to enhance energy efficiency, development of alternative materials to substitute costly critical metals, mitigating risks in handling of hydrogen and bringing it down to a level as low as practicable, and development of infrastructure for hydrogen handling must be enhanced to bring hydrogen to the energy market and to make it accessible and affordable to industries at large.

Hydrogen economics is not only electrochemistry but systems engineering as well. If the load curve is unstable, the material degradation accelerates. If the ramping profile is aggressive, the life of the electrolyser reduces fast. If electricity pricing is poorly structured, Levelized Cost of Hydrogen will rise exponentially. Very little improvement (say 1 or 2%) in stack energy efficiency, cannot compensate for a plant operating at capacity of as low as 35%. In other words, electrochemistry sets the ceiling, and the load curves determine the floor. Hydrogen economics is therefore more about systems engineering consisting of Renewable intermittency, PPA (Power Purchase Agreement) structure, Curtailment management, Storage buffering and Grid Tariff design.

Hydrogen is versatile. It is a fuel for combustion, an energy carrier, a catalyst to enhance productivity, a feedstock for many chemicals, and energy storage. There is growing global awareness of global warming and how clean hydrogen can be the right answer to address this menace. India is moving fast on bridging the gaps to bring clean hydrogen to market and mitigating the risks involved in handling it to a level as low as practicable, under its National Hydrogen Mission.

Conclusion

Green hydrogen will play a pivotal role in decarbonising the planet and could be the best alternative to fossil fuels, as it emits no carbon during combustion and is abundantly available in the universe in compounded form. It can be used as an energy carrier, an energy storage medium, a transportation fuel, an effective catalyst, and as a feedstock for the manufacture of many useful chemicals.

It has been reported that free hydrogen is available from underground reservoirs in the Pyrenees and the Alps mountain range in Europe, but this has yet to be demonstrated. It has been opined by geoscientific experts that hydrocarbons in underground reservoirs have gone through a natural geochemical reformation process forming hydrogen, which might have migrated and settled in a favourable reservoir rock. But it has yet to be demonstrated at scale.

More than 95% of hydrogen is produced globally from fossil fuels, called grey hydrogen. Grey hydrogen emits the maximum amount of carbon dioxide among other types, blue and green hydrogen. Grey hydrogen can be converted to blue hydrogen, a cleaner form of hydrogen, through carbon dioxide capture and then utilising it or storing it. The cleanest form of hydrogen, called green hydrogen, can be produced through an electrolytic process fed with renewable power, achieving purity as high as 99.999%, suitable for a hydrogen fuel cell. H_2 can also be produced from biomass, either through biogenic or thermogenic processes, but these processes are low-yielding and slow, in addition to being less energy efficient and logistically cumbersome in biomass handling.

The electrolytic process, which splits water into hydrogen and oxygen, is compact and clean, efficient, and yields the highest degree of purity. This is the reason why the electrolytic process is the most wanted process all over the world to produce clean hydrogen. But the cost of production through the electrolytic process is very high, 5 to 6 times higher than natural gas.

Because of its high diffusivity and flammability, handling of hydrogen requires extra care and a separate infrastructure. However, hydrogen can be handled with existing natural gas infrastructure, provided that the blending ratio does not exceed the upper limit of 20% at pressures below 80 bar, which otherwise can cause hydrogen embrittlement in piping material, leading to fracture. Up to 20% blending of hydrogen with natural gas is being practiced by many global industries, including India. Exceeding this limit may cause hydrogen diffusion through valves and joints, in addition to embrittlement.

Clean hydrogen demand is rising sharply for energy and non-energy uses, primarily due to decarbonisation campaigns worldwide. Hydrocarbon refineries and ammonia industries consume the maximum hydrogen. Manufacturing industries such as steel and cement, which emit high carbon during production, are also adopting hydrogen reduction process to reduce CO_2 emissions.

Almost all of the world's current H_2 production, more than 95 Mt/yr, is coming from fossils and is used for non-energy purposes. Global hydrogen production is expected to increase significantly by 2050, with forecasts ranging from 500 to 660 million tonnes per year. This growth is driven by the need to decarbonise industrial sectors, particularly hard-to-abate industries such as steel, cement, and heavy-duty transportation.

As predicted by forecasting agencies, non-energy uses of hydrogen are expected to continue to grow slowly until the mid-2030s, declining thereafter to current levels by the mid-century, with falling demand mainly associated with the decline in oil product demand and the associated use of hydrogen in refineries. Substantial growth in hydrogen demand will come from its use for energy, either directly or through ammonia and e-fuels derived from hydrogen. As per the estimates, by 2050, only 30% of the global hydrogen supply will be used for non-energy purposes. 39% will be used directly for energy, while 31% will be converted to ammonia or e-fuel for energy end users.

Low-emissions hydrogen production projects, called blue hydrogen, have gone from just a handful of demonstrations in 2021 to more than 200 committed investments, with projects increasing in number and scale, reflecting the importance of hydrogen for climate goals, energy security, and industrial competitiveness. Nevertheless, hydrogen growth has not met all of the expectations raised at the start of the decade and remains uneven. Uncertainties about costs, infrastructure readiness, and evolving regulatory frameworks present barriers to faster deployment.

Demand for grey hydrogen is projected to decline as demand for clean hydrogen rises and the costs of green hydrogen eventually become more competitive, expected by mid-thirties. According to McKinsey's *Global Energy Prospective-2023*, by 2050, clean hydrogen demand could account for up to 90% (375 Mtpa) of total hydrogen demand, with about 10%

(40 Mtpa) met by grey hydrogen in the most expected further-accelerated scenario.

At present, total hydrogen demand in India is 6 mtpa, most of which, as high as 95%, is consumed in fertilisers and refineries. As per the report by CEEW, hydrogen consumption in the fertiliser, refining, and petrochemical industries is anticipated to drive India's total hydrogen demand to 11.9 mtpa by 2030 from the level of 5.6 mtpa in 2020. India plans to produce 5 mtpa of green hydrogen through the electrolytic process by 2030, sourced from renewable energy, with an investment of Rs. 8 lakh crores and the installation of 125 GW of solar and wind power. It is estimated to provide employment to 6 lakh of workers in its complete value chain.

Global installed water electrolyser capacity reached 2 gigawatts (GW) in 2024. China now accounts for 65% of global installed capacity and capacity that has reached a final investment decision. China is also home to nearly 60% of global electrolyser manufacturing capacity. China's existing manufacturing capacity of 20 GW per year is significantly above current demand, which was around 2 GW in 2024. The cost of manufacturing and installing an electrolyser outside of China in 2024 was USD 2,000 to 2,600 per kilowatt (kW), compared with USD 600 to 1,200 per kW for electrolysers manufactured and installed in China, according to the IEA 2025 report.

Renewable power and electrolyser hardware are major cost drivers of green hydrogen, with limited scope for dramatic price drops before 2030 in India. Infrastructure, water scarcity (each kilogram of green hydrogen requires 9–20 litres of ultra-pure water), and supply chain risks pose substantial scaling barriers in India. As of late 2025, the true cost of green hydrogen production in India remains in the \$3.7–6/kg range. Globally, less than 1 million tonnes per year of green hydrogen projects actually exist in operation. Renewable power accounts for around 65–70% of the green hydrogen cost. The RTC cost of green power, including asset oversizing, battery or pumped hydro storage, and grid integration, is Rs. 4–5.1/kWh. This RTC price directly sets a floor for green hydrogen cost, because electrolysers cannot operate economically on intermittent supply alone.

The second cost pillar is electrolyser hardware. Electrolysis technologies primarily use either alkaline or PEM systems. These require precious metals such as iridium, platinum (PEM), titanium, and a specific polymer as electrolyte, or nickel-based electrodes (alkaline), all of which set a natural cost floor. Balance of plant components—rectifier, transformer, water treatment, and cooling systems—significantly increase the total cost. The global average capital cost of an electrolyser in 2025 ranges from \$500–800 per kw for PEM and \$400–700 per kw for advanced alkaline systems. Research from IEA, IRENA, and CEEW indicates that further cost reductions for electrolysers by 2030 are limited to 20–30% below current levels, primarily through supply chain and process optimisation. Material constraints and technology maturity may not let the cost fall sharply.

Electrolysers operate at 65–70% energy efficiency. Incremental improvements in efficiency of just 5–10% are likely over a decade due to fundamental limits in electrochemistry, offering a 5–7% reduction in overall hydrogen production costs. A solid oxide electrolyser is reported to be 85% efficient, but it is yet to be fully developed and commercialised. Electrolysis is a material-heavy, precision-engineered device, constrained by electrochemistry, not just manufacturing volume and automation. Scaling production may deliver 10–15% cost savings, but a dramatic price drop is unrealistic.

India has positioned green hydrogen as a flagship solution for its energy transition, targeting 5 million tonnes of annual production by 2030, up from the current level of a few kilotons, and universal adoption across sectors such as steel, fertilisers, refining, and transport. While the rapid rise in solar and wind power production can well address the ambition, electrolyser manufacturing will need significant impetus.

The objective of climate policy is solely focused on how much CO_2 reduction can be achieved for each dollar spent. At present, the CO_2 abatement cost of green hydrogen is very high, ranging from \$70 to \$ 175 per tonne. This makes sense for wealthier nations, which can absorb the risk of scaling up and have abundant capital and infrastructure. As technology develops, demand grows, and the cost reduces.

Solid Oxide and Anion Exchange batteries are emerging technologies that are expected to deliver high efficiency at lower cost, because they do not need costly electrodes and electrolytes. By the mid-1930s, these technologies are expected to be well developed and cost-effective, reducing the cost of green hydrogen to levels comparable to grey hydrogen and even lower.

The Global North, which has flooded the atmosphere with CO_2 during rapid industrialisation (out of a hundred molecules of CO_2 in the air, more than three-fourths belong to it), must shoulder the responsibility for developing economies of scale. For the Global South, like India, where resources are scarce, and the need for fast and sizeable returns is paramount, priority may be given to proven CO_2 abatement techniques such as renewables, blue hydrogen, gasification, biofuel, grid modernisation, efficiency enhancement, and recycling, which can deliver the large CO_2 reductions at lower cost. Blue hydrogen production by CCUS technology is the least-cost low-carbon hydrogen and can be produced through SMR and coal gasification pathways. India owns 9.5% of the global coal reserve, and through the gasification process, it can be best utilised in power as well as hydrogen production. For India, adopting green hydrogen after economies of scale and innovation have made the green hydrogen cost truly competitive, shall be a better proposition. India should focus more on R&D works to excel in catalytic electrolysis technology and other technologies like biogenic and thermogenic, to produce hydrogen.

Non-energy uses of green hydrogen in hard-to-abate industries such as refineries, fertilisers, petrochemicals, and others can be prioritised. We can consider adopting hydrogen for energy applications at a later stage, once electrolyser technology has fully developed and matured. The use of hydrogen in the energy sector requires substantial systemic changes—ranging from modifications in transport vehicles and large-scale adoption of fuel cells (a technology still evolving), to the establishment of robust and safe infrastructure for its storage, transportation, and distribution. It will take time to adopt green hydrogen in the transport sector, whether through fuel cell or combustion, because it requires a major transformation in technology and manufacturing to make it usable, accessible, and affordable to the public at large.

There are many R&D efforts underway across many fields of science and technology, such as electrochemistry for electrolysers, CO_2 hydrogenation for carbon removal, materials science for H_2 storage and transportation, and many others. Complementary inventions in nano-catalysis, plasmonic catalysis, and advanced materials are enabling more durable, efficient, and cost-effective energy solutions. Technological advancements in green hydrogen production, CCU, e-fuel production, and AI-driven operations are redefining the global energy landscape, paving the way for sustainable, resilient, and carbon-neutral industries.

Coal did not disappear with the arrival of oil, nor did oil vanish with the rise of natural gas and renewables. Similarly, hydrogen will not replace all existing fossil fuels, but certainly it will reduce and reverse its consumption to a minimum, where the environment will remain protected from global warming.

Clean hydrogen shall remain the main focus, as it is always preferred, because preventing CO_2 emissions is better than investing efforts and economy in mitigating them. Now, it is no longer a conviction but a realisation that clean hydrogen will be a sustainable fuel for the future.

Looking at the current and future scenario of hydrogen energy, technological advancement and utilization, commitment to meet the climate target, and dependency on huge energy imports, it is suggested that

1. More focus be given to low-carbon blue hydrogen production through carbon capture, coal gasification, thermogenic and biogenic pathways from biomass.

   At present, the CO_2 abatement cost of green hydrogen is very high. Electrolytic technology, which produces green hydrogen, has yet to develop fully and mature to achieve higher energy efficiency and lower cost. Critical metals availability for PEM electrolysers will remain a big constraint for a long till any alternative comes up. India should avoid large investment towards green hydrogen at this stage and wait for the electrolytic technology to develop further to be more efficient and cost-effective, which is likely to happen from 2030 onwards. However, it must continue pursuing green hydrogen production at a pace.
2. Impetus in renewable energy production must continue to tap the optimum solar and wind potential. Energy storage, grid integration, and round-the-clock power availability system should be ensured through AI-aided technology.
3. India is facing twin challenges of increasing energy demand and meeting the climate target. In such a situation, nuclear power shall be a big tool. Nuclear power generation should be augmented from the present 0.8 to 10% in phases. India is rich in thorium reserves, but very poor in fission-grade uranium; therefore, breeder reactors should be pushed more to use thorium in nuclear power generation. Solid Oxide Electrolyser works at as high as 90% energy efficiency at high temperature steam, which can be obtained from nuclear power plants.
4. India should give more emphasis to R&D works, particularly in emerging Solid Oxide and Anion Exchange electrolyser technology, which are more efficient and economical than other types in use, and find an alternative to costly materials used in electrodes and electrolytes in PEM electrolysers, to make it more energy efficient and cost-effective. It should try to excel in technology more than to exceed in numbers at this stage in the field of electrolytic hydrogen production.

Glossary

(Hydrogen - Energy of the Future)

Activation Energy - Activation energy is the minimum amount of energy required to initiate a chemical reaction by breaking bonds in reactant molecules, acting as a potential energy barrier that must be overcome for products to form. It determines the reaction rate; higher activation energy results in slower reactions.

ATR - Autothermal Reforming (ATR), an efficient, compact method for hydrogen production that combines steam methane reforming and partial oxidation in a single reactor, using oxygen/air and steam to convert hydrocarbons into syngas (H_2 + CO). By using exothermic oxidation to fuel endothermic reforming, it eliminates the need for external heating, offering faster startups, lower capital costs, and superior CO_2 capture capability for blue hydrogen.

ATJ Process - The Alcohol-to-Jet (ATJ) process is a commercially proven pathway for producing Sustainable Aviation Fuel (SAF) by converting alcohols (ethanol, isobutanol) derived from biomass, waste gases, or sugar/starch into jet fuel. It involves dehydrating, oligomerising, and hydrogenating alcohol to create synthetic hydrocarbons, with up to 50% blending allowed.

Anthropogenic emission - Anthropogenic emissions are greenhouse gases (GHGs) and other pollutants released into the atmosphere primarily from human activities like burning fossil fuels, deforestation, industrial processes, and agriculture, significantly driving climate change by trapping heat and altering the natural system.

Blue Hydrogen - Blue hydrogen is produced from natural gas using Steam Methane Reforming (SMR) or Autothermal Reforming (ATR), paired with Carbon Capture and Storage (CCS) to trap CO_2 emissions. As a lower-carbon alternative to grey hydrogen, it aids industrial decarbonisation, but it is not carbon-free, as some emissions remain.

Biogenic pathway - A pathway to convert one compound to another through biological process with application of enzymes and bacteria. Fermentation is a biological pathway to convert sugar to ethanol.

Bagasse - Bagasse is the fibrous, pulpy residue remaining after crushing sugarcane or similar plants (like agave) for their juice. It is a sustainable, renewable, and biodegradable agricultural byproduct, widely used as an eco-friendly alternative to plastic and paper in food packaging, as well as a renewable bioenergy source.

Batch Process - A batch reactor is a closed, non-steady-state vessel where reactants are added at the beginning, mixed, and allowed to react for a set time before the product is removed. It operates without continuous inflow or outflow, making it ideal for small-scale production, pharmaceuticals, and studying reaction kinetics. These reactors offer high flexibility for manufacturing multiple products.

Bioconversion - Bioconversion (or biotransformation) is the eco-friendly process of using microorganisms, enzymes, or biological agents to convert organic waste, plant matter, or agricultural byproducts into value-added products like biofuels (biogas, ethanol), fertilisers, and industrial chemicals.

Biogas - Biogas is a renewable fuel produced through the anaerobic digestion (decomposition without oxygen) of organic waste materials like cow dung, sewage, and food waste. It primarily consists of 55–65% methane and 35–45% carbon dioxide, making it a versatile, clean-burning alternative for cooking, heating, and generating electricity.

Biosphere - The biosphere is the global ecosystem encompassing all living organisms (biotic) and their interactions with the atmosphere, hydrosphere, and lithosphere (abiotic). It is the thin, 20-km thick, life-sustaining zone on Earth extending from deep ocean trenches to high mountain tops, including all ecosystems and biomes.

Breakeven cost - The breakeven cost (or point) is the production level where total revenue equals total costs, resulting in zero profit or loss.

Carnot Cycle Efficiency - Carnot cycle efficiency is the maximum theoretical efficiency any heat engine can achieve operating between hot (T_h) and cold (T_c) reservoirs, defined as

$$\eta = 1 - \frac{T_C}{T_H} \textbf{ (using Kelvin).}$$

Carbon Cycle - The carbon cycle is the process that moves carbon between plants, animals, and microbes; minerals in the Earth; and the atmosphere. Carbon is the fourth most abundant element in the universe. With its ability to form complex molecules such as DNA and proteins, carbon makes life on Earth possible.

Carbon Neutral - Making or resulting in no net release of carbon dioxide into the atmosphere, especially as a result of carbon offsetting. Net zero is similar in principle to carbon neutrality, but is expanded in scale. To achieve net zero means to go beyond the removal of just carbon emissions. Net zero refers to all greenhouse gases being emitted into the atmosphere, such as methane (CH_4), nitrous oxide (N_2O), and other hydrofluorocarbons.

Capacity Factor - A capacity factor is the ratio of a power plant's actual energy output over a period to its maximum potential output if it operated at full nameplate capacity 24/7, expressed as a percentage. It measures reliability and usage, with high factors indicating constant, efficient operation, while lower factors reflect intermittent, renewable, or demand-driven sources.

Clean Coal Technology - Clean coal technology comprises a set of advanced methods designed to reduce the environmental impact of coal-fired power generation by minimising pollutants like sulphur dioxide, nitrogen oxides, and carbon dioxide. Key techniques include Carbon Capture and Storage (CCS/CCUS), High Efficiency Low Emission (HELE) systems (e.g., supercritical plants), IGCC (Integrated Gasification Combined Cycle) power plant, and coal washing to remove impurities.

Clean Hydrogen - Clean hydrogen is a fuel produced with near-zero greenhouse gas emissions, primarily through electrolysis powered by renewable energy (green hydrogen) or natural gas paired with carbon capture and storage (blue hydrogen).

CCUS - Carbon Capture, Utilisation, and Storage(CCUS) is a critical suite of technologies designed to mitigate climate change by capturing CO_2 emissions from high-emission industrial sources (power plants, factories) before they enter the atmosphere. Captured CO_2 is either transported to be stored permanently deep underground or used for manufacturing products like fuels and construction materials.

Decarbonisation - Decarbonisation is the process of systematically reducing or eliminating human-caused greenhouse gas emissions—particularly carbon dioxide and methane—from economic activities like energy production, transport, and manufacturing. It aims to achieve net-zero emissions by transitioning to renewable energy, enhancing efficiency, and using carbon capture technologies to mitigate climate change.

Deforestation - Deforestation is the large-scale, permanent removal of forests for agricultural expansion, urbanisation, and logging, with nearly 90% driven by agriculture. This process fuels climate change by releasing stored carbon dioxide, destroys biodiversity, disrupts water cycles, and causes soil erosion. It currently accounts for significant habitat loss and environmental degradation.

DRI Process - The Direct Reduced Iron (DRI) process (or sponge iron process) reduces iron ore (pellets or lumps) to metallic iron in the solid state using reducing gases (H_2, CO) or coal at 1,000–1,050°C. It removes oxygen from the ore without melting it, creating a porous structure used primarily in Electric Arc Furnaces (EAF) to produce high-quality steel.

Electrolytic Cell - An electrolytic cell is an electrochemical device that uses an external source of electrical energy (DC) to drive a non-spontaneous chemical reaction, a process known as electrolysis. It converts electrical energy into chemical energy to decompose compounds or plate metals.

Electrolysis - Electrolysis is a process that uses direct electric current (DC) to drive a non-spontaneous chemical reaction, decomposing ionic compounds into their elements, water decomposing into hydrogen and oxygen.

Electrochemistry - Electrochemistry is the branch of chemistry that studies the relationship between electrical energy and chemical changes, focusing on electron transfer during oxidation-reduction (redox) reactions.

Electrode Potential - Electrode potential is the electromotive force (EMF) or voltage developed at the interface between an electrode (typically a metal) and its electrolyte solution, measuring a substance's tendency to undergo reduction (gain electrons) or oxidation (lose electrons).

Equilibrium Voltage - Equilibrium voltage (or potential) is the specific electrical potential difference across a membrane or in an electrochemical cell where the net flow of ions or charge is zero, balancing chemical driving forces with electrical forces. It is also termed as reversible voltage and depends upon temperature and pressure.

Economy of Scale - Economies of scale are cost advantages reaped by companies when production becomes efficient, where the cost per unit decreases as total output increases. By spreading fixed costs (e.g., machinery, rent) over a larger volume of goods, businesses lower their average costs, improving profitability and allowing for competitive pricing.

Ecological Balance - Ecological balance is a state of dynamic equilibrium within an ecosystem where biodiversity—including humans, plants, animals, and microorganisms—remains relatively stable through co-existence and interaction with their environment. It is a sustainable, functional state where population numbers, nutrient cycles, and energy flows are maintained. It is also termed as a balance between the atmosphere, hydrosphere, and biosphere.

Energy Density - Energy density is the amount of energy stored within a given system, substance, or region of space per unit volume (volumetric) or mass (gravimetric). It measures a material's or device's ability to store energy, with higher density meaning more energy in a smaller size or weight. Common units include *Wh/L* or *Wh/kg*.

Enzyme - Enzymes are specialised protein molecules that act as biological catalysts, accelerating essential chemical reactions (such as digestion and metabolism) by lowering activation energy without being consumed.

Entropy - Entropy is a fundamental concept in physics and thermodynamics, representing a measure of molecular disorder, randomness, or the unavailability of a system's energy to do work. It quantifies the tendency of energy to disperse and spread out, with higher entropy indicating more disorder and lower energy quality.

Enthalpy - Enthalpy (H) is a thermodynamic state function representing the total heat content of a system, defined as the sum of its internal energy (U) and the product of pressure (P) and volume (V), expressed as $H = U + PV$. Change in Enthalpy is measured as the sum of the change in Gibbs Free Energy and the change in Entropy at constant temperature.

Fuel Cell - A fuel cell is an electrochemical device that directly converts the chemical energy of a fuel (typically hydrogen) and an oxidising agent (usually oxygen) into electricity through redox reactions. Unlike batteries, fuel cells provide a continuous, high-efficiency power source as long as a constant supply of fuel and oxygen is maintained, producing only water and heat as byproducts.

Fermentation - Fermentation is an Error! Hyperlink reference not valid. where microorganisms like bacteria and yeast convert carbohydrates (sugars/ starch) into acids, gases, or alcohol to produce energy.

FT Process - The Fischer-Tropsch (FT) process is a catalytic chemical reaction that converts syngas—a mixture of carbon monoxide and hydrogen into high-quality liquid hydrocarbons, such as synthetic fuel, diesel, and wax. Developed in the 1920s, this technology is used for gas-to-liquids (GTL) and coal-to-liquids (CTL) conversion, operating at 150–350°C with iron or cobalt catalysts.

Galvanic Cell - A galvanic (or voltaic) cell is an electrochemical cell that converts chemical energy from spontaneous redox reactions into electrical energy.

Greenhouse Gas - Greenhouse gases (GHGs) are atmospheric gases—primarily carbon dioxide, methane, nitrous oxide, and water vapour—that trap heat from the Earth's surface, acting as a blanket that warms the planet.

Global Warming Potential - Error! Hyperlink reference not valid. (GWP) is a metric used to measure and compare the relative heat-trapping potency of different greenhouse gases in the atmosphere over a specific time horizon (commonly 100 years), relative to carbon dioxide. It quantifies the cumulative radiative forcing impact of a unit mass of a gas, with assigned a reference value of 1.

Grey Hydrogen - Grey hydrogen is hydrogen produced from fossil fuels, primarily natural gas (methane) via steam methane reforming (SMR) or coal gasification, where the resulting carbon dioxide emissions are released into the atmosphere. It is currently the most common, cost-effective form of hydrogen, accounting for ~95% of global production.

Green Hydrogen - Green hydrogen is a clean, zero-carbon fuel produced by splitting water into hydrogen and oxygen via electrolysis, powered exclusively by renewable energy sources like wind or solar. It is critical for decarbonising hard-to-abate sectors, such as steel, shipping, and chemicals. The well-to-wheel emission of CO_2 should be typically below 2 kg CO_2 equivalent per kg of green H_2 as per the DOE standard of the US, which is followed by India.

Gasification - Gasification is a thermochemical process that converts carbonaceous materials—such as coal, biomass, or waste—into a combustible gas mixture called syngas (synthesis gas), primarily hydrogen and carbon monoxide. Operating at high temperatures (700°C) with limited oxygen/steam, it breaks down materials without combustion. The resulting syngas is used for electricity generation, liquid fuels, or chemical production.

Gibbs Free Energy - Gibbs Free Energy (G) is the energy in a system available to do useful work at constant temperature and pressure, essentially the maximum energy that can be converted into work, with its change (ΔG) determining if a reaction or process will happen spontaneously. A negative ΔG means the process is spontaneous (spontaneous), while a positive ΔG means it's non-spontaneous, requiring energy input, and $\Delta G=0$ indicates equilibrium.

Hydrogen Economy - A hydrogen economy envisions a future energy system using hydrogen as a clean, versatile fuel to replace fossil fuels, aiming for decarbonisation in transport, industry, and power generation.

Hydrocracking - Hydrocracking is a two-stage petroleum refining process that combines catalytic cracking and hydrogenation to convert heavy, high-boiling-point feedstocks into high-value, light products like gasoline, jet fuel, and diesel. Operating at high temperatures and pressures, it uses bifunctional catalysts (acidic and metallic) to break large hydrocarbon chains while simultaneously adding hydrogen to improve product quality and remove sulphur/nitrogen.

Hydrotreating - Hydrotreating is a critical petroleum refining process that removes contaminants like sulphur, nitrogen, and heavy metals from hydrocarbon streams using hydrogen and catalysts at high temperatures and pressures. It produces cleaner, high-quality fuels—such as ultra-low-sulphur diesel (ULSD)—and protects downstream catalyst units. Key reactions include desulfurisation, denitrogenation, and hydrogenation of aromatics.

Hybrid System - Hybrid energy systems in India combine multiple power sources—primarily solar, wind, and battery storage—to enhance grid stability, increase efficiency, and maximise land usage. Under national policy, these systems (e.g., wind-solar hybrid) require at least one source to be at least 25% of the other's capacity. They are crucial for bridging the gap between fluctuating renewable output and consistent demand.

Hard to Abate - "Hard to abate" refers to industrial sectors—typically steel, cement, chemicals, shipping, and aviation—that find it technically or economically challenging to reduce their greenhouse gas emissions. These sectors, responsible for roughly 40% of global direct CO_2 emissions, rely on high-temperature heat or chemical processes that are difficult to decarbonise with current technology and high-cost alternatives.

HEFA Process - The HEFA process (Hydroprocessed Esters and Fatty Acids) is a leading method to create Sustainable Aviation Fuel (SAF) by upgrading waste oils, fats, and non-food vegetable oils into clean, drop-in jet fuel, involving step like hydrodeoxygenation (removing oxygen), cracking, and isomerisation to meet jet fuel specifications, and is currently the most mature and commercially available pathway for aviation decarbonisation, blending up to 50% with fossil jet fuel.

Hydrogen Carrier - Hydrogen carriers are materials or compounds designed to store and transport hydrogen in a chemically bound state, offering safer, higher-density alternatives to high-pressure or cryogenic liquid hydrogen. Key carriers include liquid organic hydrogen carriers (LOHCs), ammonia, methanol, and metal hydrides, which facilitate storage at ambient conditions and leverage existing infrastructure.

IPCCC - The Intergovernmental Panel on Climate Change (IPCC) is the UN body for assessing science related to climate change, established in 1988 by the WMO and UNEP. With 195 member countries, it provides policymakers with regular, rigorous scientific assessments on climate change impacts, risks, and mitigation strategies, based on peer-reviewed literature.

Life cycle of a compound - The life cycle of a chemical compound involves a cradle-to-grave approach, tracking its journey from resource extraction, manufacturing, and use to disposal or recycling. This process, also known as a chemical life cycle, involves mapping material and energy flows and assessing environmental impacts like toxicity and degradation, as described by the Sustainability Directory.

Levelised cost of Hydrogen - The Levelised Cost of Hydrogen (LCOH) is the average, discounted cost per kilogram of hydrogen produced over a project's lifetime, covering capital (capex), operating expenses (opex), and financing. It acts as a key metric for benchmarking green hydrogen's competitiveness against fossil-fuel-based methods.

NDC – NDC (Nationally Determined Contribution) is a non-binding national plan to reduce greenhouse gas emissions and adapt to climate change under the Paris Agreement, with the purpose to o achieve the goal of limiting global warming to well below 2°C (preferably 1.5°C), updated every 5 years with increasing ambition.

Net-Zero Emission - Net-zero emission refers to achieving a balance where the amount of greenhouse gases (GHG) produced by human activity equals the amount removed from the atmosphere. It is not absolute zero, but rather cutting emissions to a minimum, with remaining emissions balanced by carbon sinks. Global targets aim for net-zero by 2050 to limit warming to 1.5°C, with India targeting 2070.

Ocean Acidification - Ocean acidification is the ongoing decrease in the pH of the Earth's oceans, caused by the uptake of atmospheric carbon dioxide (CO_2) from human activities like burning fossil fuels. This process has increased ocean acidity by 30% since the industrial era. It creates carbonic acid in seawater, reducing available carbonate ions needed by marine organisms to build shells and skeletons, which threatens ecosystems.

Ozone Layer - The ozone layer is a thin shield of gas (triatomic oxygen, O_3) in the stratosphere, roughly 15-35 km above Earth, that protects life by absorbing 97-99% of the sun's harmful ultraviolet (UV) radiation. Human-produced substances like CFCs caused significant depletion, notably the "hole" over Antarctica. Thanks to the Montreal Protocol, this layer is healing and expected to recover.

Oxy Fuel - Oxy-combustion (or oxy-fuel combustion) is a technique that burns fuel in nearly pure oxygen mixed with recycled flue gas instead of air, resulting in a flue gas primarily composed of CO_2 and water vapour. This process simplifies CCS by producing high-purity CO_2 while increasing combustion efficiency and reducing NO_x emissions.

Overpotential - Overpotential is the extra voltage applied to an electrochemical cell, beyond the theoretical equilibrium potential required to drive a reaction at a specific rate. It represents energy lost as energy due to kinetic barriers and is categorised into activation, concentration, and ohmic types.

PV Module - A PV (photovoltaic) module is an assembly of interconnected silicon solar cells that convert sunlight into direct current (DC) electricity. Serving as the building block of solar power systems, these modules are typically encapsulated between tempered glass and a protective back sheet, framed in aluminium, and used in residential, commercial, and utility-scale projects to generate renewable energy. A PV module is a single, factory-assembled unit of interconnected solar cells protected by a frame and glass.

PV Panel - Photovoltaic (PV) panels are used to produce electricity directly from sunlight. PV panels consist of a number of individual cells connected together to produce electricity of a desired voltage. Photovoltaic panels are inherently DC devices. A PV panel is a broader term, often referring to one or more modules installed together to generate electricity.

Rare Earth Elements - Rare earth elements (REEs) are a group of 17 metallic elements—the 15 lanthanides (lanthanum to lutetium) plus scandium and yttrium—essential for high-tech, green energy, and defence applications. Despite their name, they are relatively abundant in the Earth's crust but difficult to find in concentrated, economically exploitable deposits.

Reformation - Reformation is a chemical process that rearranges the molecular structure of hydrocarbons to improve their properties, primarily upgrading low-octane naphtha into high-octane components for gasoline. Using heat and catalysts (often platinum), this technique breaks down or converts molecules to enhance combustion. It is also used to produce hydrogen gas.

SMR Process - The SMR (Steam Methane Reforming) process is the primary industrial method for producing hydrogen, involving reacting natural gas (methane) with high-temperature steam over a catalyst to create a mixture of hydrogen and carbon monoxide (syngas), followed by a Water Gas Shift (WGS) reaction to convert CO into H_2.

Substrate - In chemistry, a substrate is the specific molecule, compound, or species that undergoes a chemical transformation or is acted upon by a catalyst or enzyme to form a product. It is the foundation molecule in organic synthesis or the substance targeted by enzymes in biochemical reactions, often binding at an active site to trigger a reaction.

Sponge Iron - Sponge iron, or Direct Reduced Iron (DRI), is a metallic product created by reducing iron ore (lumps, pellets, or fines) with coal or natural gas at 800 to 1200°C, without melting it. This porous, high-purity (84–95% Fe) material serves as a crucial, high-quality substitute for scrap in electric arc and induction furnaces.

SAF Production - Sustainable Aviation Fuel (SAF) is a cleaner alternative to conventional jet fuel, made from sustainable resources like waste oils, municipal waste, and non-food crops, capable of reducing CO_2 emissions. It is a vital, immediate solution for decarbonising aviation, compatible with existing aircraft engines and infrastructure, though currently limited by high costs and production scale.

Thermogenic pathway – A pathway to convert one compound to another compound by the application of heat energy. Gasification is a thermogenic pathway where fossil or biomass is converted to hydrogen and other compounds with the application of heat.

Thermodynamic - Thermodynamics is the branch of physics and chemistry concerned with heat, work, temperature, and their relation to energy and physical properties of matter. It governs energy conversion, describing how thermal energy is converted to/from other forms, with key principles including the 4 laws of thermodynamics and the concepts of entropy and equilibrium.

Thauer Limit - The Thauer limit refers to a theoretical maximum yield in dark fermentative biohydrogen (H_2) production, which dictates that a maximum of 4 moles of H_2 can be produced per mole of glucose consumed. This limitation is derived from thermodynamic restrictions in single-organism metabolism, specifically during glycolysis, where sugar is converted into acetate, CO_2, and H_2.

Thermodynamic Voltage - Thermodynamic voltage (or reversible voltage) is the maximum theoretical electromotive force (E) of an electrochemical cell, determined by the Gibbs Free Energy change (delta G) of the reaction, where E = - deltaG / nF. It represents the equilibrium voltage at a given temperature and composition without losses, calculated via the Nernst equation.

UNFCCC - The UNFCCC (United Nations Framework Convention on Climate Change) is the main global treaty and UN body for coordinating international action against climate change, aiming to stabilise greenhouse gases to prevent dangerous human interference with the climate system. Adopted in 1992, it serves as the parent treaty for key agreements like the Kyoto Protocol and the Paris Agreement, with its secretariat based in Bonn, Germany, managing yearly Conference of the Parties (COP) meetings to assess progress and negotiate solutions.

Well-to-Wheel - Well-to-Wheel (WTW) analysis is a comprehensive method for evaluating the total energy consumption and greenhouse gas emissions of vehicle fuel throughout its entire lifecycle, from energy production ("well") to vehicle operation ("wheel")

Yield Percent - Yield in a chemical conversion process is the amount of desired product obtained relative to the maximum possible (theoretical) amount, usually expressed as a percentage. It serves as a critical KPI for process efficiency, representing how effectively raw materials are converted into the final product while accounting for losses or side reactions. Theoretical yield is the maximum product possible, based on the stoichiometry of the limiting reactant.

www.ingramcontent.com/pod-product-compliance
Lightning Source LLC
LaVergne TN
LVHW051134210926
848482LV00006B/312